Efficient Extreme Ultraviolet Mirror Design

An FDTD approach

IOP Series in Advances in Optics, Photonics and Optoelectronics

SERIES EDITOR

Professor Rajpal S Sirohi Consultant Scientist

About the Editor

Rajpal S Sirohi is currently working as a faculty member in the Department of Physics, Alabama A&M University, Huntsville, Alabama (USA). Prior to this, he was a consultant scientist at the Indian Institute of Science Bangalore, and before that he was chair professor in the Department of Physics, Tezpur University, Assam. During 2000–11, he was academic administrator, being vice chancellor to a couple of universities and the director of the Indian Institute of Technology Delhi. He is the recipient of many international and national awards and the author of more than 400 papers. Dr Sirohi is involved with research concerning optical metrology, optical instrumentation, holography, and speckle phenomenon.

About the series

Optics, photonics and optoelectronics are enabling technologies in many branches of science, engineering, medicine and agriculture. These technologies have reshaped our outlook, our way of interaction with each other and brought people closer. They help us to understand many phenomena better and provide a deeper insight in the functioning of nature. Further, these technologies themselves are evolving at a rapid rate. Their applications encompass very large spatial scales from nanometers to astronomical and a very large temporal range from picoseconds to billions of years. The series on the advances on optics, photonics and optoelectronics aims at covering topics that are of interest to both academia and industry. Some of the topics that the books in the series will cover include bio-photonics and medical imaging, devices, electromagnetics, fiber optics, information storage, instrumentation, light sources, CCD and CMOS imagers, metamaterials, optical metrology, optical networks, photovoltaics, freeform optics and its evaluation, singular optics, cryptography and sensors.

About IOP ebooks

The authors are encouraged to take advantage of the features made possible by electronic publication to enhance the reader experience through the use of colour, animation and video, and incorporating supplementary files in their work.

Do you have an idea of a book you'd like to explore?
For further information and details of submitting book proposals see iopscience.org/books or contact Ashley Gasque on Ashley.gasque@iop.org.

Efficient Extreme Ultraviolet Mirror Design

An FDTD approach

Yen-Min Lee
ASML Holding

IOP Publishing, Bristol, UK

ISBN 978-0-7503-2652-0 (ebook)
ISBN 978-0-7503-2650-6 (print)
ISBN 978-0-7503-2653-7 (myPrint)
ISBN 978-0-7503-2651-3 (mobi)

DOI 10.1088/978-0-7503-2652-0

Version: 20210901

IOP ebooks

British Library Cataloguing-in-Publication Data: A catalogue record for this book is available from the British Library.

Published by IOP Publishing, wholly owned by The Institute of Physics, London

IOP Publishing, Temple Circus, Temple Way, Bristol, BS1 6HG, UK

US Office: IOP Publishing, Inc., 190 North Independence Mall West, Suite 601, Philadelphia, PA 19106, USA

To my students:

I understand, most of time, you don't choose the things you want to do.
They choose you.

Please keep in mind, the best way to get yourself prepared is to realize there is no such thing as well-prepared.

Contents

Preface

As I am sitting in front of Scheldt and staring at the grand scenery mirrored from the calm water, I can't help but think of the many uses of mirrors. With retrospect to the history of humanity, the earliest man made mirrors can be dated to around the 6th millennium BC, we can imagine that the need to manufacture these primitive mirrors was out of basic grooming. Nowadays, instead of only serving as a grooming tool, mirrors are almost always present no matter where we are. On the one hand, they help us in our daily mundane lives. On the other hand, they play an important role to push for advanced developments in technology. Most optical devices are composed of mirrors. Can you imagine ophthalmic surgery equipment without a laser? Trying to think how inconvenient your life would be if you are using a smart phone without a built-in camera? How about a satellite without solar panels? In addition, new important applications to use a mirror can also be found in the cutting-edge semiconductor industry—extreme ultraviolet (EUV) lithography.

EUV lithography is expected to be the silver lining to the death of Moore's Law. When I started writing this book, EUV lithography was on the way to becoming the star of next-generation chip processing technology. The commercial EUV lithography systems are maturing and starting to ship to global semiconductor manufacturing companies for high volume maunfactoring. We can anticipate that for the coming years, it is expected to speed up the development of 5G technologies, artificial intelligence, edge computing, more complex virtual reality applications, the Internet of thing, etc.

As I started my PhD career in engineering, I noticed micro-lithography technology would be key to economic growth and development in the near future. It started to dominate our technology development. I was fascinated by the elegant craft of micro-lithography technology. Bear in mind that since the year 2000, this has been one of the major technologies, which enables 3C industrial business to grow and CPU technology to advance by leaps and bounds. As a result, these progresses facilitate not only computational science to advance rapidly but also mobile devices to gain extensive popularity. Through serendipity, during the second semester of my PhD career, I began my EUV lithography research with strong support from ASML, an EUV mirror was one of the major topics. Nevertheless, a EUV mirror did not yet exist. Therefore, most of the research was highly reliant on computer-aided modeling. Owing to the mesoscopic length scale of an EUV mirror, it is always a tradeoff between accuracy and computational time. After digging into this specialty for a year, I saw a bunch of research that utilized a over simplified model in calculations. In addition, for rigorous simulation, it is still easily to get bogged down even if we are using a super computer because the simulation consumes gazillions of memory. As an example, take one of my cases during my PhD, it takes years to finish a rigorous simulation for a complete EUV mirror with ~50 CPUs to run in parallel. To that end, I was looking for an alternative favorable for this research. Eventually, the efficient finite-difference time-domain (FDTD) approach was developed and I believe it will have a significant contribution to EUV mirror design. The efficient FDTD approach described in this book intensely reduces the computational resources required, and the computational accuracy is comparable to that of rigorous simulation. Detailed implementation of this methodology together with simulation examples will be shared throughout this book.

I wrote this book primarily for EUV industry professionals, especially, and individuals who've also been investigating large-scale problems or near-field scattering problems will find it useful. Also, this book can be a good starting point for readers who have not had a chance to learn about photo-lithography and the FDTD method. This book starts with an introduction to photo-lithography, general readers can expect to understand not only photo-lithography itself but also how integrated circuits were made through photo-lithography. Moreover, this book describes how an EUV mirror is made. After reading this section, readers will have knowledge concerning mirror composition and its fabrication process. This book also explains how do construct an efficient FDTD method and how it works based on the foundation of the FDTD method. For those who don't know about the FDTD method, this section will help to understand the basics of the FDTD method. For readers already familiar with the FDTD method, this book offers an efficient approach to doing computation using the FDTD method. Finally, there will be EUV mirror calculation cases demonstrated in the last chapter.

There are a lot of people who inspired me when writing this book, I was encouraged by many academics, schoolmates, friends, colleagues and even my family. It would be impossible for me to name them all and acknowledge them one by one, but I would like to give my sincere thanks to every single one of them.

Yen-Min Lee
Antwerpen, September 2019

Author biography

Yen-Min Lee

Yen-Min Lee, PhD, has worked at ASML since 2014, with the responsibility of new algorithm design for lithography tools. Throughout his career in ASML, he has worked in the USA, in the Netherlands and in Taiwan on different products and projects. Before joining ASML and before finishing his PhD, in 2012, Dr Lee had experience as a technical intern at Siemens (Princeton, NJ, USA) for a medical ultrasound transducers project. At Siemens, he participated in ultrasound based elastography equipment development. In particular, he focused on new algorithm development for medical imaging applications based on a linear elasticity model. Dr Lee received his PhD from National Taiwan University in 2013. His main research theme was large-scale EM simulations, inverse problems for optical and medical applications, optimization processes for thin film applications, especially for thin film transistor liquid crystal display (TFT-LCD) and EUV mirror, and lithography modeling.

Abbreviations and symbols

2D	Two-dimensional
λ	Wavelength
k_1	Proportional coefficient of the lithography process complexity
M	Metal layer
R	Resolution
Hg	Mercury
W	Tungsten
ADI	After Development Inspection
Al	Aluminum
ARC	Antireflection Coating
ArF	Argon Fluoride
BARC	Bottom Antireflection Coating
BEOL	Back End of Line
BW	Bandwidth
CD	Critical Dimension
CMP	Chemical Mechanical Polish
CMOS	Complementary Metal-oxide Semiconductor
CPU	Central Processing Unit
Cu	Copper
CVD	Chemical Vapor Deposition
CW	Central Wavelength
DOI	Defect of Interest
DSA	Directed Self-assembly
DUV	Deep Ultraviolet
EP	Edge Placement
EPE	Edge Placement Error
EUV	Extreme Ultraviolet
FEOL	Front End of Line
GOX	Gate Oxide
H	Hexahethyldisilazane
HCE	Hot Carrier Effect
HVM	High Volume Manufacturing
IC	Integrated Circuit
ILD	Inter Layer Dielectric
IMD	Inter-metal Dielectric
KrF	Krypton Fluoride
LDD	Lightly Doped Drain
LER	Line Edge Roughness
LPCVD	Low-pressure Chemical Vapor Deposition
ML	Multilayer
MOS	Metal-oxide Semiconductor
MOSFET	Metal-oxide Semiconductor Field Effect Transistor
MP	Multiple Patterning
NA	Numerical Aperture
NI	Nanoimprint
NMOS	N-type MOS
NUV	Near Ultraviolet

PMOS	P-type MOS
PR	Photoresist
PSG	Phosphosilicate Glass
SAC	Self-aligned Contact
STI	Shallow Trench Isolation
TARC	Top-layer Antireflection Coating
TSMC	Taiwan Semiconductor Manufacturing Company
UV	Ultraviolet
VIA	Vertical Interconnect Access
VLSI	Very Large-scale Integration
1D	One-dimensional
3D	Three-dimensional
θ	Angle
θ_{i}	Angle of incidence
θ_{r}	Angle of reflection
θ_{t}	Angle of refraction
θ_{vacuum}	Angle in vacuum
α	Factor of neutral plane of in-plane distortion
β	Phase change
Π	Periodicity of the periodic structure
Δx	Deformation along x-direction
Δx_{err}	Image error in x-direction
Δy	Deformation along y-direction
Δz	Deformation along z-direction
Δz_{err}	Image error in z-direction
T_{mirror}	Thickness of the mirror
AltPSM	Alternating Phase-Shift Photomask
AttPSM	Attenuated Phase-Shift Photomask
BIM	Binary Intensity Photomask
C	Carbon
CAD	Computer-aided Design
Cl_2	Chlorine
Cr	Chromium
$\vec{E}$	Electric field vector
$\vec{E}_{\mathrm{i}}$	Electric field vector of the incident wave
$\vec{E}_{\mathrm{r}}$	Electric field vector of the reflected wave
$\vec{E}_{\mathrm{t}}$	Electric field vector of the transmitted wave
EM	Electromagnetic
F_2	Fluorine
$\vec{H}$	Magnetic field vector
$\vec{H}_{\mathrm{i}}$	Magnetic field vector of the incident wave
$\vec{H}_{\mathrm{r}}$	Magnetic field vector of the reflected wave
$\vec{H}_{\mathrm{t}}$	Magnetic field vector of the transmitted wave
IPD	In-Plane Distortion
I_{i}	Illuminated intensity
I_{r}	Reflected intensity
k_{a1}	Extinction coefficient of absorber 1 in AttPSM
k_{a2}	Extinction coefficient of absorber 2 in AttPSM
$\vec{K}$	Bloch wave vector
LTEM	Low Thermal Expansion Material

m	Required etching level of the designed AltPSM
$M_{Mo/Si}$	Characteristic matrix corresponding to one Mo/Si period
Mo	Molybdenum
Mo_{eq}	Equivalent Mo layer
$\tilde{n}$	Complex refractive index
$\tilde{n}_{Mo}$	Complex refractive index of Mo
$\tilde{n}_{Si}$	Complex refractive index of Si
N	Pair number
NGL	Next-generation Lithography
OPD	Out-of-Plane Distortion
O_2	Oxygen
PHC	Photonic Crystal
PVF	Peak-to-Valley Flatness
r	Reflection coefficient
RAM	Random Access Memory
Rb	Rubidium
R_{top}	Total reflection coefficient of the topmost surface
Ru	Ruthenium
$\vec{s}$	Unit vector of the direction of propagation
SEMI	Semiconductor Equipment and Material International
Si_{eq}	Equivalent Si layer
Si	Silicon
Sr	Strontium
t^p	Transmission coefficient of p-polarization
t^s	Transmission coefficient of s-polarization
T	Thickness
Ta	Tantalum
T_{a1}	Thickness of absorber 1 in AttPSM
T_{a2}	Thickness of absorber 2 in AttPSM
TE	Transverse-electric
TEC	Thermal Expansion Coefficient
Ti	Titanium
Tm	Required etching depth of the designed AltPSM
TM	Transverse-magnetic
$u(\vec{r})$	Bloch function
x_r	Image reduction ratio of the EUV lithography system
ψ_x	Real scale in x-direction
ψ_y	Real scale in y-direction
ψ_z	Real scale in z-direction
Δ_{comp}	Grid numbers along one axis
Δ_{ref}	Referenced grid numbers along one axis
μ_0	Permeability of free space
ε_0	Permittivity of free space
ε_r	Relative permittivity
σ	Conductivity
σ_e	Electric conductivities of PML
σ_m	Magnetic conductivity of PML
ξ	Factor related to FLOPs
τ	Time step
v	Wave speed

η	Intrinsic impedance of normal illumination
η'	Intrinsic impedance of oblique illumination
η_{eq}	Equivalent intrinsic impedance of the entire ML
γ	Characteristic propagation constant
ABC	Absorption boundary condition
BW	Bandwidth
$\overline{B}$	Magnetic flux density
CW	Central wavelength
CXRO	Center for x-ray optics
$\overline{D}$	Electric flux density
DIM	Differential method
e_{rms}	RMS error
FDTD	Finite-difference time-domain
FEM	Finite-element method
FLOPs	Floating point operations per second
GB	Gigabyte
$\overline{J}$	Current density
k_x	Wave number in *x*-direction
k_y	Wave number in *y*-direction
k_z	Wave number in *z*-direction
n_d	Number of dimensions
n_e	Number of unknowns
$\widetilde{n}_{eq}$	Equivalent refractive index
N_c	Mesh number in computational domain
N_x	Mesh number in *x*-direction
N_y	Mesh number in *y*-direction
N_z	Mesh number in *z*-direction
N_{data}	Number of total sample points.
OPL	Optical path length
PBC	Periodic boundary condition
PEC	Perfect electric conductors
PML	Perfect matched layer
RCWA	Rigorous coupled wave analysis
RMSE	Root mean square error
R_{CXRO}	Reflectivity from the CXRO solver,
$R_{Fresnel}$	Reflectivity calculated from the Fresnel's equations based solver
t	Time
t_{comp}	Computational time
t_{ref}	Reference computational time
T'	Equivalent length
WGM	Waveguide method
Z	Impedance of the finite transmission line
Z_{eq}	Equivalent impedance
Z_{in}	Input impedance
Z_0	Impedance of the vacuum
Z_{PML}	Impedance of PML
Z'	Impedance of oblique illumination
A	Intensity at the center of an EUV light source
F	A factor of full width at half maximum
FWHM	Full width at half maximum

KPI	Key performance indicator
LPPS	Laser produced plasma light source
MPI	Message passing interface
ML_{eq}	Equivalent ML stack
TaN_{eq}	Equivalent TaN layer

Chapter 1

Introduction to optical lithography

Lithography technologies have courageously advanced down to smaller critical dimension (CD), which is a core technology for the fabrication of integrated circuits (IC) that form the backbone of the global communications and advanced computation capabilities that have transformed modern life over the last half-century. According to the Rayleigh criterion, the most intuitive way to shrink the CD size is to apply a new light source with shorter wavelength (λ) rather than use the existing ultraviolet (UV) light, that is, the light wavelength shorter than 193 nm or other lithographic techniques should be applied. This chapter will start with the basic working principle of optical lithography, the following section will be the evolutionary history of this technology. From that point, the interrelation with IC manufacturing will be described. At the end of this chapter, extreme ultraviolet (EUV) lithography, which is the next-generation lithography to enable the continued extension of Moore's law, will be described.

1.1 Principle of optical lithography

Optical lithography is also called photolithography or sometimes just lithography. Lithography refers to the process invented in 1796 by Alloys Senefelder [1], where patterns of desired designs were transferred on to a base substrate, mostly using photomasks (or reticle). One of the important roles that optical lithography used to play is to enable the continued extension of Moore's law [2], that is, the number of transistors in a dense IC doubles every two years. The simplified version of this law states that the speeds and overall processing power for a central processing unit (CPU) will double every two years. The simplified version of optical lithography is a process to delicately transfer the designed layout information from the photomask onto the silicon wafer through light propagations. Metaphorically, the process to transfer the photomask layout onto the wafer is pretty much the same with the process to transfer the camera image onto the photographic film. Figure 1.1 is a schematic sketch of the optics of one of the common optical lithography systems,

doi:10.1088/978-0-7503-2652-0ch1

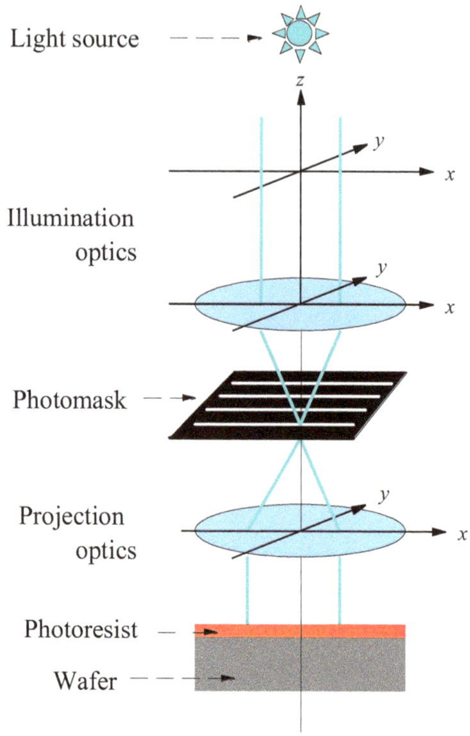

Figure 1.1. Schematic illustration of the optics of a projection printing lithography system.

this configuration is the optics of the so-called projection printing system [3]. It is very common that, in a typical optical lithography system, the lights are designed to pass through all the optical elements all the way to the wafer. However, this isn't the case for an EUV lithography system since all its optics have to be reflective. We will have a brief introduction to EUV lithography in section 1.4. Regardless, the basic concept to transfer the photomask layout onto the wafer is the same.

As shown in figure 1.1, the optics of an optical lithography system typically consists of five segments, which are a light source, illumination optics, photomask, projection optics and the silicon wafer. The UV light source utilizes 193 nm central wavelength is the mainstream and was used in general optical lithography systems before EUV was selected as a new volume production tool. This UV light source irradiates UV lights along the $-z$ direction. Once these UV lights tend to move across the illumination optics, they will be converged by the condenser in the illumination optics before moving through. Afterward, the converged UV lights need to pass through a carrier, called a photomask. A photomask consists of a plate that is transparent to the UV light source, on which the geometric layouts are defined by using opaque elements on its backside. These layouts are what designers want to see on the wafer. As a results, once the UV lights pass through the transparent regions, the layouts can then be propagated towards the projection optics together with the UV light waves. The lights in the projection optics will again be focused by the

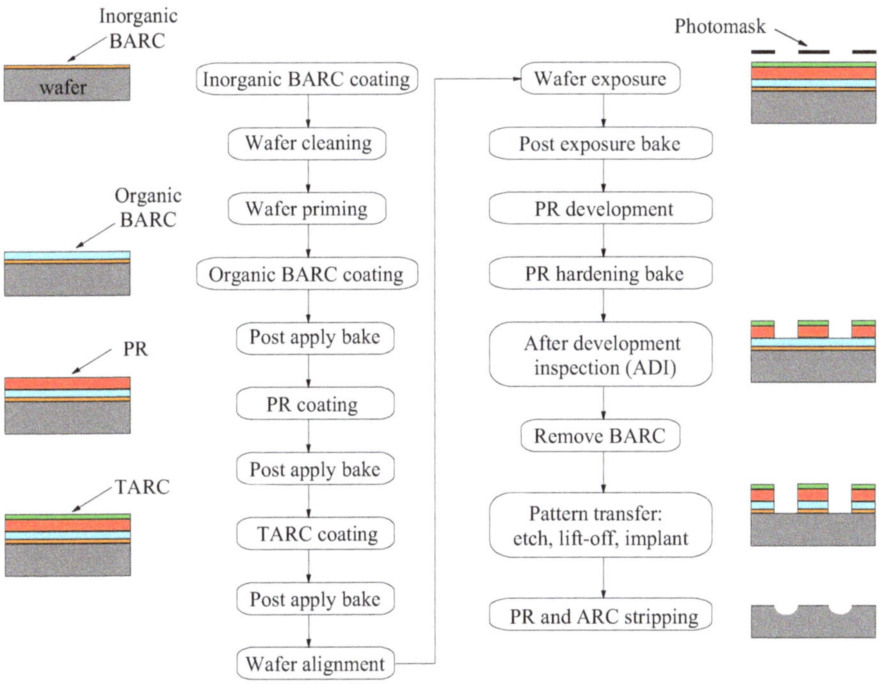

Figure 1.2. Simplified block diagram of generic optical lithography process steps.

condenser. After the lights pass through projection optics, the photoresist (PR) layer on the wafer is exposed to UV radiation, which modifies the molecular structure followed by a change in the solubility of the material, while placing the photomask of layouts. The above steps briefly describe how the optics of the optical lithography system transfers the layouts from the photomask to the wafer, this process is also called exposure or pattering.

Nevertheless, in order to realize the layouts on the wafer, there are more steps to be performed, before- and after-exposure, in a general optical lithography process, as shown in figure 1.2 [4]. It is known that the UV lights passing through the PR film will reach the wafer, so they are expected to reflect off the wafer. Under this circumstance, the reflected lights interfere with illumination lights to form standing waves in the direction normal to the wafer [5]. These strong standing waves introduce a corrugated profile in the PR film thus result in poor CD with line edge roughness (LER). To minimize these standing waves, the PR film will be sandwiched between different antireflection coatings (ARCs). In a general optical lithography process, an inorganic ARC will be deposited first on the wafer using chemical vapor deposition (CVD) methods. Afterward, the wafer will be cleaned and primed to enhance the adhesion property. One of the most common primers used for this purpose is hexamethyldisilazane (HMDS) [6]. Then, an organic ARC will be spin-coated onto them and baked at high temperatures to remove moisture and to cross-link the films. Both of these ARCs are so-called bottom antireflection coatings (BARCs) [7] since they will be located below the PR film. The next step of the lithography process is to form a uniform PR film on organic BARC through spin coating. During PR coating, the centrifugal force of the wafer spinning with several thousand rounds per minute distributes a few milliliter of PR over the wafer surface area. After the coating process is completed, the PR film contains a remaining solvent concentration depending on the PR, the solvent and the PR film thickness. The solvent in the PR film could induce a photomask contamination, foaming inside the PR, a weak PR adhesion, etc. To reduce the residual solvent in the PR film, another post apply bake (or soft bake) is performed after finishing PR coating. After well-preparing the PR film, a so-called top-layer antireflection coating (TARC) [8] is applied above the PR film. As with the previous layers, a soft bake is also applied to this TARC layer. This TARC is used to reduce the reflections of the incident light at the air/PR interface. At this stage, the wafer is well-prepared and needs to align with the photomask.

Afterward, the wafer exposure described in figure 1.1 is proceed. After another soft bake, the wafer is immersed in a developer solution to realize the layouts on the PR. Developer solutions are typically aqueous and dissolve away areas of the PR exposed to UV lights. Moreover, a PR hardening bake (or hard bake) is proceed to stabilize the printed layouts and provide optimum performance for etching. The next step is to perform after development inspection (ADI) through optical metrology or electron microscope. ADI is very important in the optical lithography process because it allows engineers to catch defects of interest (DOI) earlier so that the wafer can be reworked to remove PR and redo the PR patterning. BARC removal and etching are carried out subsequently to complete the entire lithography process.

1.2 Evolution and history

The precision of transistor gate dimensions is determined by the resolution of the optical lithography process and it is a critical path in complementary metal-oxide semiconductor (CMOS) transistors downscaling. The resolution of the lithography system is defined by how sharp the minimum CD can be distinguished. The resolution is determined by Rayleigh's law [9]:

$$R = \frac{k_1 \lambda}{\mathrm{NA}} \tag{1.1}$$

where R is the resolution, λ is the wavelength of applied light, k_1 represents the proportional coefficient of the lithography process complexity, and NA is the numerical aperture. For that reason, the resolution of the lithography system can be improved by decreasing λ, k_1 and increasing the NA. Based on equation (1.1), the direct method to achieve smaller feature sizes is to apply light sources with shorter wavelength in optical lithography systems.

The evolution of optical lithography in the history of humanity can be seen in table 1.1 [10]. Over the past few decades, the used wavelength of optical lithography has varied from the blue spectrum at 436 nm to the EUV at 13.5 nm, which is a reduction in wavelength of 96.9%. By mixing additional techniques, such as immersion and multiple patterning, higher NA is also achieved. As a result, the corresponding CD has shrunk from 7000 nm to 7 nm, which is a 99.9% decrease.

Table 1.1. Evolution of optical lithography methodologies.

Methodology	Initial CD	First recorded year of usage
Contact	7000	1962
Proximity	3000	1972
Projection	2000	1973
Electron beam	500	1976
X-ray	300	1978
G-line	1250	1978
I-line	800	1985
DUV (248 nm)	450	1986
DUV (193 nm)	150	1996
DUV (157 nm)	100	1998
DUV (193 nm)	65	2007
DUV immersion (193 nm)	45	2009
DUV immersion with MP (193 nm)	10	2016
EUV (13.5 nm)	7	2018

Before 1997, the common light source of the optical lithography systems was vaporized mercury (Hg) arc gas discharge lamps. Such a light source produces a spectrum centered at around 400 nm wavelength with different spectral lines at 436 nm, 405 nm, and 365 nm, which are also termed the G-line, H-line, and I-line, respectively. In this connection, the earliest optical lithography methodology was recorded in the 1960s, the methodology utilizes the photomask to direct contact with the PR on the wafer. Therefore, this methodology is so-called contact lithography. At the beginning, the reachable CD using contact lithograph was 7 μm.

A decade later, proximity lithography was born to replace the contact lithography, and a resolution of 3 μm was recorded when first utilizing this methodology. Technically, what proximity lithography improves upon contact lithography is the photomask induced defects during direct PR contact. Although at the cost of complex optics, projection lithography was dominated after a year with the benefit of converging the de-magnified lights onto the wafer. In order to adapt to the downscaling, alternative forms of lithography methodologies had come about that rely on the use of different source, such as electron beam lithography and x-ray lithography. Both of methodologies provide higher resolution than previous technologies. However, both of them have fatal flaws. For electron beam lithography, the slow throughput destroyed the possibility of this methodology to become a high volume manufacturing (HVM) tool. For x-ray lithography, the photomask membrane must be made very thin and internal stress could result in distortion of the absorber, this disadvantage brings down its overall appeal.

In view of this, a Hg lamp light source based optical lithography then prevailed from 1978 to 1985. Along with the progression of IC technology, the wavelength of applied light in the optical lithography process was reduced from the G-line to I-line in the near ultraviolet (NUV) region. Afterward, in the development of the semiconductor industry, the continuous downscaling road map also requested higher throughput to achieve the goals to increase the density transistors in the chip and to reduce the production costs. However, the Hg lamp as the light source could not meet these industry's needs. Therefore, the methodology with 248 nm deep ultraviolet (DUV) generated by an excited dimer krypton fluoride (KrF) laser was born for HVM with CD larger than 150 nm. Later, the argon fluoride (ArF) laser at a wavelength of 193 nm was developed to fabricate IC for CD down to 150 nm and above. Owing to the resolution-enhanced immersion technology, the 193 nm ArF lithography technology can meet the 45 nm technology node requirements [11].

Immersion lithography technology is filled with purified water between the lens of projection and the substrate, which has a higher refractive index than air. In this way, the light beam is focused to obtain a high resolution for smaller nano-scaled patterning [12]. Therefore, the NA of the projection prism can be greatly increased to obtain the high resolution. After the further improvement of lithography technology, the 193 nm ArF immersion with multiple patterning is the best developed technique for 10 nm node [13]. Multiple patterning (MP) is a type of resolution enhancement method in lithography to increase the feature density. Unlike the shrinking reticle patterns to smaller sizes, the multiple patterning technique is based on exposure to different reticles to offset each other [14], which is then followed by wafer etching process steps.

At present, Moore's law seems to be extended by utilizing an EUV light source in optical lithography that enables the 7 nm node. EUV lithography has been invested in for more than a decade by semiconductor industrial leaders such as Taiwan Semiconductor Manufacturing Company, Limited (TSMC), Intel and Samsung. It is predicted that they might start to use EUV lithography as a next-generation tool for high-volume IC manufacturing before 2024.

1.3 Interrelation with IC manufacturing

Since the 1960s, optical lithography has been used by the semiconductor industry as one of the crucial manufacturing processes for making transistors and ICs of very large-scale integration (VLSI) [10]. Nowadays, because of improvement in optical resolution, IC devices are widely applied everywhere and the technology node of an IC device CD has shrunk down to one ten-thousandth, compared to the dimension in 1960. Based on the above statements, readers should notice that the resolution of optical lithography influences not only the yield of the transferred layout profiles on the PR layer but also the size of the wafer to be used in IC manufacturing, because it governs the minimum feature size, which is also called the critical CD. Since the function of optical lithography is to transfer the layouts from the photomask to the wafer, the overall liability that comes along with CD control is to fabricate one- and two-dimensional (2D) features in the right places, this is called edge placement (EP). Due to the complication of the entire optical lithography process and optical diffraction effect, usually the patterned features undergo mis-alignment or shrinkage issues, the phenomenon is the so-called edge placement error (EPE) [15]. EPE can be broadly categorized into the following three types

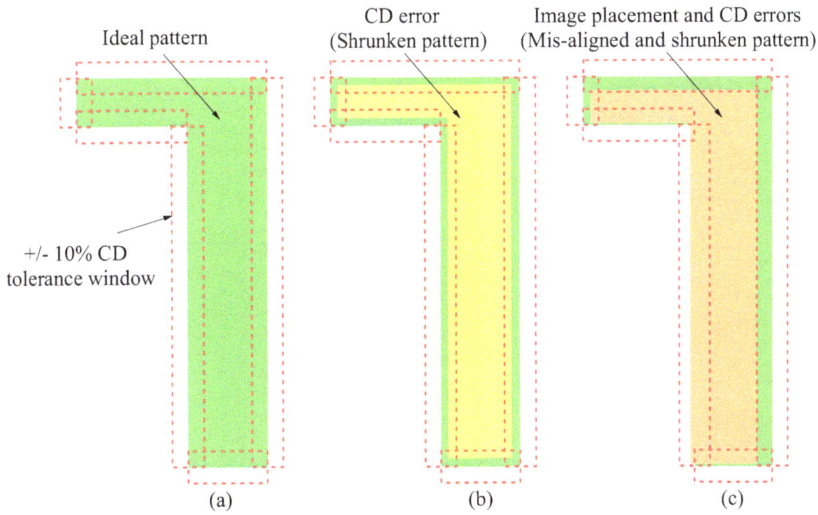

Figure 1.3. Goal of optical lithography: top views of edge placements of a Γ shape feature, in which the red dashed lines sketch out to a 10% CD tolerance window. The green feature shown in (a) stands for the ideal pattern, the yellow feature shown in (b) stands for the shrunken pattern, and the orange feature shown in (c) stands for the pattern that is suffering from mis-aligned and shrunken effects.

(1) Line edge roughness (LER) [16]: when variations in the width of a feature occur quickly over the length of the feature, this variation is called LER.
(2) CD error [17]: the difference between EPE of the right edge and left edge of a feature, as shown in figure 1.3(b).
(3) Image placement error: the average error in center position of a feature over a length of interest, as shown in figure 1.3(c).

EPE examples can be seen in figure 1.3, assuming the green Γ shape feature is a 2D feature that chip designers would like to print on the wafer. The Γ shape feature undergoes different allocations as shown in figures 1.3(a)–(c), on which the red dash lines outline the 10% tolerance windows of each CD. The CD error induced by pattern shrinkage is plotted as a yellow Γ shape in figure 1.3(b). It is obvious that both the left and right sides of the yellow feature have a difference compared to the green pattern due to contraction. The combination of both image placement and CD errors can be seen in figure 1.3(c), we can see the orange Γ shape suffers not only image mis-alignment but also feature shrinkage at certain areas. Regardless, the pattern is considered to be qualified as long as it is located within the tolerance window.

To sum up, EPE is the difference between the intended and the printed features of an IC layout. If one or more EPE issues crop up in the production flow, the IC device is subject to shorts or poor yields, which could cause the entire chip to fail. For this reason, to fabricate the advanced ICs, the challenges of the optical lithography process have the highest priority, which have to be considered and resolved among all the IC manufacturing processes. Figure 1.4 briefly depicts the

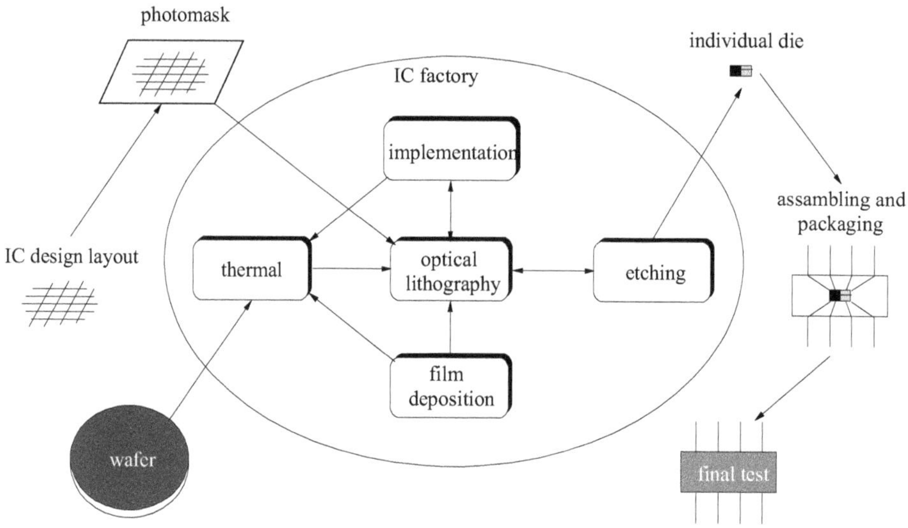

Figure 1.4. Major IC fabrication process steps.

major IC fabrication process steps, readers are able to see the correlation between optical lithography and the entire IC fabrication process. Please note that in reality the entire IC fabrication process must experience hundreds of sequential steps combining physical and chemical reactions alternatively on the silicon wafer. Normally, the entire process takes six to eight weeks.

As shown in figure 1.4, IC design layouts must be printed on the photomask before it goes into the IC factory. In the IC factory, there are generally five sub-processes [18], to enable a bare wafer to turns out with several dies, which are thermal process, implantation process, optical lithography process, etching process and film deposition process. Below are the brief introductions to sub-processes out of the scope of optical lithography:

Thermal process: This process operates at high temperature, the temperature usually reaches up to 800°C which is higher than the melting point of aluminum (Al). This is the front-end process to prepare the wafer for the optical lithography process step, basically it takes place in a high temperature furnace.

Implementation process: The use of accelerated ions to dope and modify material properties of specific wafer regions, generally called ion implantation. The ion implanters dope channel structures at certain doping density to set the threshold conditions for transistor switching, enabling the practical fabrication of CMOS transistors that are now the dominant form of IC devices.

Deposition process: This process step is used to add thin layers to the wafer surface, these layers can be of a different material or a different microstructure or composition of the same material. The different layers help in defining the various components of the MOSFET and in obtaining a functional device e.g. the passivation layer helps in electrically isolating the metal contacts to the source, drain and gate.

Etching process: This is the process of using strong acid or etchant to cut into the unprotected parts of a metal surface to create a design. Typically, dry etching and wet etching are used in the IC factory. Dry etching refers to any of the methods of etching that use gas instead of chemical etchants, this is capable of producing critical geometries that are very small. Wet etching uses chemicals where a batch of wafers is dipped into a highly concentrated pool of acid and the exposed areas of the wafer are etched away, this is good and fairly cheap and capable of processing many wafers quickly. In the photolithography process etching is used to remove unnecessary PR. Mostly, it is used during metallization to remove unnecessary metal surface for making contact between devices.

In order to keep up with the HVM demand of IC manufacturing, all the sub-processes need to be applied over and over again so that the required layouts can be fabricated in large batches on every single wafer. To be more specific, the layouts will need to be repeatedly fabricated in different areas on a single wafer with extreme high repeatability. After completing all sub-processes over and over again in the IC factory, the wafer now contains multiple identical ICs to individual dies. Then a wafer with up to thousands of circuits is cut into rectangular pieces through die sawing, each of them is called a die, each containing one copy of the fabricated layouts. These dies are actually one of the sub-functional circuits that compose an IC chip. Dies are then attached onto the lead frame using the epoxy adhesive or solder. After curing the die attach paste, the gold, copper, aluminum wires are bonded between dies. In order to separate these bounded dies and to form the leads into the correct shape and position, lead frame strips need to go through a trim/form/singulation machine. Afterward, these semiconductor products are packed for a final electrical test.

IC is very small in size, which is composed of a semiconductor wafer on which millions of small resistors, capacitors and transistors are fabricated. The process flows to construct a CMOS from the bottom to the top can be spit into two process lines, which are the front end of line (FEOL) and back end of line (BEOL). The former is meant to build the transistor function, while the latter is meant for metallization and electric connection. In IC chips, the base substrate is a silicon wafer. On top of the substrate, the IC chip is stacked up by many overlapping layers that contain sub-functional circuits, and all components are constructed from a specific combination of these layers. The contexts below express both FEOL and BEOL with illustrations:

(1) **Front end of line (FEOL)**

- Shallow trench isolation (STI): STI is a methodology to isolate the active regions of the transistors. As shown in figure 1.5(a), a p-type wafer is served as a substrate and placed into a high temperature furnace in order to form a pad oxide layer on top of it. Then, an SiN layer is deposited above the pad oxide through low-pressure chemical vapor deposition. Afterward, a PR layer will be coated for lithography purposes. The resulted chip after the lithography process can be seen in figure 1.5(b). The next step is to place this chip into the furnace to form the thin liner layers. Now, the SiO_2 can be deposited to fill the

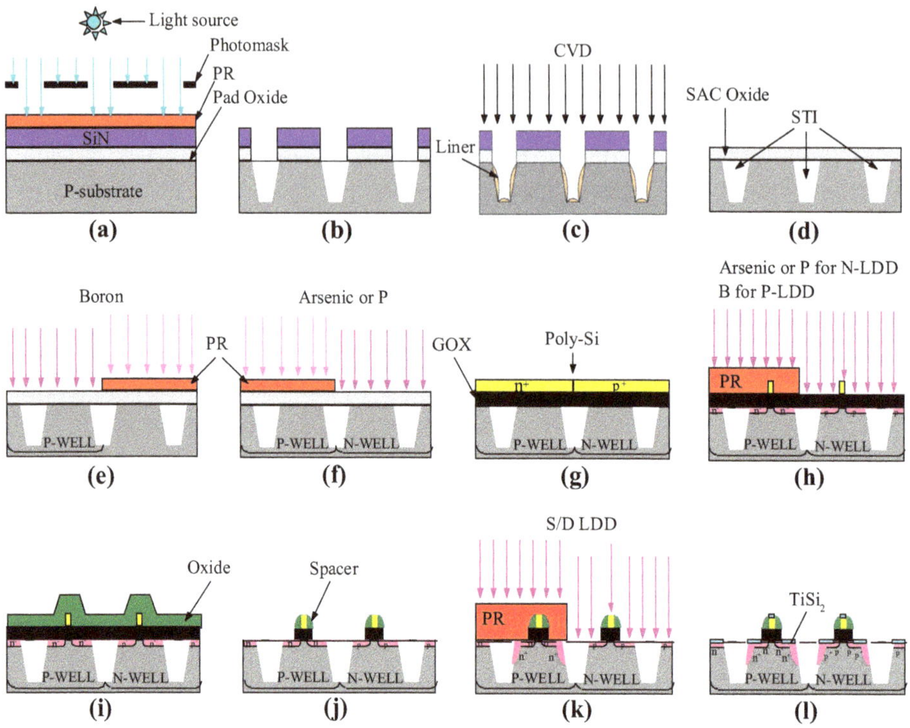

Figure 1.5. Front end of line (EBOL) process.

STI gaps, as illustrated in figure 1.5(c). The last step to complete STI is to utilize chemical mechanical polish (CMP) to rub and polish the SiN surface.

- Wells forming: in N-well technology, an n-type well is diffused on a p-type substrate, whereas in P-well it is vice-verse. The basic process is to clean the pad oxide layer through dilute hydrofluoric (HF) acid. Then, this chip is put into a high temperature furnace to form a self-aligned contact (SAC) oxide layer, as shown in figure 1.5(d). After experiencing lithography and implementation processes in both n-well and p-well regions, as shown in figures 1.5(e) and (f), this metal-oxide semiconductor (MOS) now has basic structures to integrate both NMOS and PMOS transistors over the same silicon wafer.
- Poly-silicon gate: this is the gate structure of the transistor. The first step, as shown in figure 1.5(g), is to fabricate both p+ poly-Si and n+ poly-Si above the gate oxide (GOX) layer. To prevent the hot carrier effect (HCE), both n-type and p-type lightly doped drain (LDD) are formed by using lithography, as shown in figure 1.5(h). Then, an oxide layer is deposited through CVD. Furthermore, the next step is to create spacers with anisotropic dry etching. These two steps can be seen in figures 1.5(i) and (j).

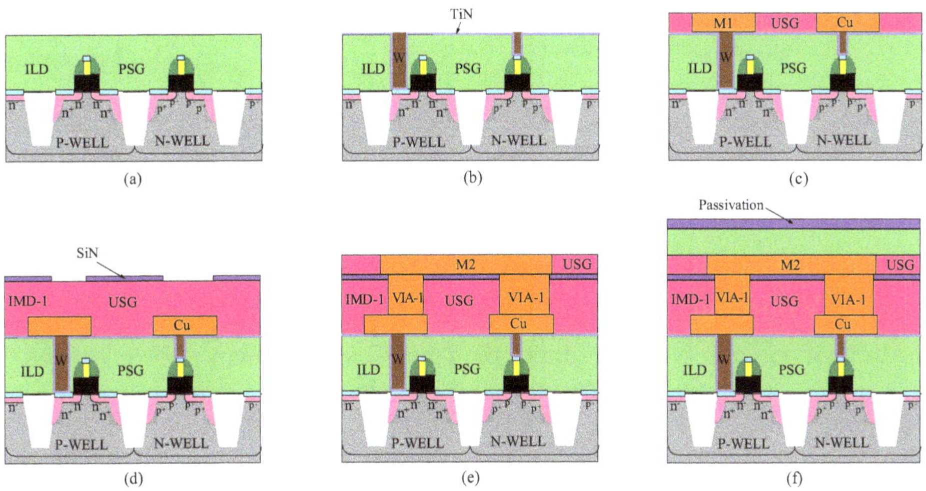

Figure 1.6. Back end of line (BEOL) process.

- Source/drain/silicide: the last two steps to complete FEOL are to construct source, drain and silicide through lithography and P/B implementation, as shown in figures 1.5(k) and (l). The function of silicide $TiSi_2$ is to enhance the adhesion between Si and the upcoming metal deposition. This way the resistivity can be decreased.

(2) **Back end of line (BEOL)**

- Inter layer dielectric (ILD) and contact holes: the ILD is used to electrically separate closely spaced interconnect lines arranged in several levels (multilevel metallization) in an IC. It must feature low dielectric constant to minimize capacitive coupling between adjacent metal lines. As shown in figure 1.6(a), the ILD is built by depositing phosphosilicate glass (PSG) above the silicide. After using CMP on its surface, the ILD is formally completed. Instead of separating electrical connections, the contact holes are tunneled by using lithography and etching for conductivity purpose. After that, the tungsten (W) is deposited into the holes as shown in figure 1.6(b).
- Metal (M), inter-metal dielectric (IMD) and vertical interconnect access (VIA): to form the first metal layer (M1), a pure undoped silicate glass (USG) layer has to be deposited above W holes and the ILD layer by using CVD. After a lithography process, the M1 is built by depositing copper (Cu). After using CMP on its surface, the M1 is formally completed, as shown in figure 1.6(c). Before fabricating the second metal layer (M2), an USG layer is deposited above M1 by using CVD to serve as a IMD-1 layer. Afterward, a SiN layer is deposited and then etched according to the etch-stop design of the succeeding IMD-2 layer, as shown in figure 1.6(d). After that, another USG layer is deposited and etched. What is going to happen then is to deposit Cu to form both

vertical interconnect access (VIA-1, between M1 and M2) and M2, as shown in figure 1.6(e). To stack up more metal layers (M3, M4, etc), we just need to repeat the process of making M2 and VIA-1.

- Passivation: in order to prevent transistors from contamination and moisturizing, a PSG layer is deposited on top of the last metal layer. Then, a SiN layer will be deposited as shown in figure 1.6(f).

1.4 The next-generation optical lithography: extreme ultraviolet (EUV) lithography

The diffraction limit, which describes the theoretical maximum resolving power of applied wavelength, foresees that optical lithography using 193 nm UV light is not able to advance Moore's law at a certain point. Technology options for next-generation lithography include EUV, electron beam, nanoimprint (NI) and directed self-assembly (DSA). All of these technologies have significant challenges that must be overcome. Regardless, EUV lithography is the most promising lithographic technology and has been chosen to enable the continued extension of Moore's law. In 2019, TSMC started to produce 7 nm chips with EUV lithography machines and it determined to go down to 5 nm by 2020. The basic concept of EUV lithography [19] is to produce a bright source of 13.5 nm radiation and use reflective optics to reduce the photomask image and transfer the pattern into resist on the wafer surface. Since most materials absorb photons in this wavelength range, reflective optics was developed and the photomask must also be a reflective photomask. Since this is a new technology, critical challenges must be overcome:

- A 13.5 nm radiation source with sufficient power and adequate bandwidth (BW): as of 2019, the highly considered EUV light source is a laser-pulsed tin plasma [20]. The basic specifications of EUV source are specified as a 13.5 nm central wavelength (CW) with 2% BW at intermediate focus, they are chosen based on multilayer (ML) mirrors design in an EUV lithography system. Besides, the output power is also an important characteristic for EUV source as the throughput is tied to the source power. To reach 120 wafers/hour, the collectable power target for EUV lithography is at least 250 W [21], while for other conventional lithography sources it is much less [22]. For example, immersion lithography light sources target 90 W, dry ArF sources 45 W, and KrF sources 40 W.

Figure 1.7 is an example to demonstrate an EUV power consumption from the EUV source illumination optics to the wafer in an EUVL system. The projection optics contains 7 Ml blank (or substrate) based reflectors, they are one photomask and six mirrors M_5–M_{10} with reflectivity assumed to be 65%. At the beginning, a 250 W in-band 13.5 nm CW EUV with 2π Steradian is excited. In general, after experiencing four mirrors in the collection optics, it remains 2% power for the projection optics. In projection optics, the light keeps weakening because of multiple reflections, optical path attenuations and BW mismatch. Eventually, the power drops to 250 mW for the photolithography process.

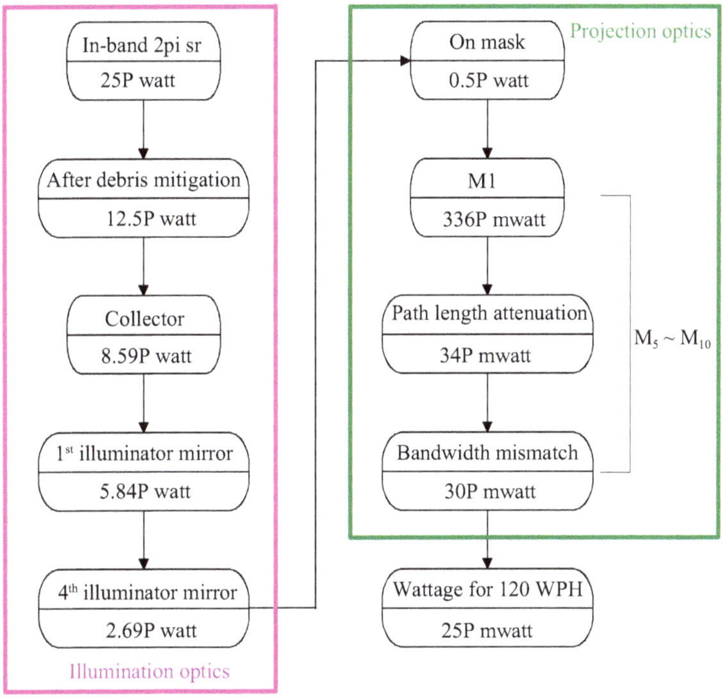

Figure 1.7. Example of EUV power degradations from illuminator to the wafer.

- A reflective optical system that works in vacuum: owing to high absorption of EUV radiation, the optical system must be accomplished in the vacuum. Since nothing transmits well in this region, the optical components must employ reflective optics that typically has losses of 35%–45% for each mirror. We illustrate an example of EUV lithography system, which can be seen in figure 1.8. As mentioned also in figure 1.7, the light degradation is caused not only by reflections but also transmission attenuations. Thus, the degradation of power start to occur before the EUV reaches the photomask. Afterward, it degrades significantly after reflecting from the photomask. In addition to photomask and illumination optics, there are typically six mirrors in the projection optics. Therefore, with a total of eleven reflections, only 25 PmW is available on the wafer. Additionally, to accurately align multiple optical axes between mirrors is also a challenge that must be accomplished by this system.

- EUV mirror technology: the EUV mirror has another set of challenges since most materials have strong absorption in the EUV region. Thus, transparent optical lenses and masks are not available, and the EUV power loses a lot after four-time (illumination optics) to seven-time (projection optics) reflections in the system. Therefore, it is critical to design the mirrors (or mask) with satisfied reflectivity for the EUV scanner. Thus far, it utilizes a set of

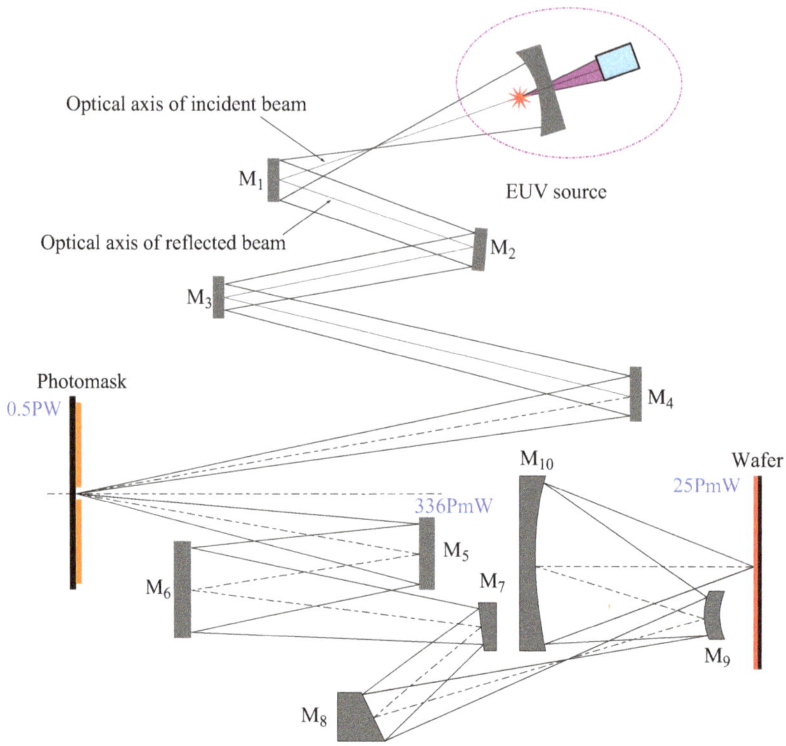

Figure 1.8. Example of EUV lithography system, which is composed of illumination optics (EUV source and M_1–M_4) and projection optics (photomask and M_6–M_{10}).

layers of silicon and tungsten to constructively reflect the EUV radiation. We will discuss this in more detail in chapter 2.

– Adequate PR for 13.5 nm radiation: since the intensity of the radiation reaching the surface of the PR is greatly reduced, the PR needs to have a high sensitivity and the ability to produce small high-resolution features while also having low LER and tight control of feature size, with reduced outgassing [23–25]. PR film thicknesses may continue to shrink with feature size, in part to avoid pattern collapse [26, 27], which may introduce new issues. Future semiconductor processes might also require several different post-processing methods [28] to meet the projected patterning requirements. While work on positive chemically amplified PR will continue, research is underway to explore other potential material candidates that satisfy projected requirements for enabling emerging lithographic technologies. These include: non-chemically amplified PR materials, inorganic, organic–inorganic hybrid resist materials, novel negative PR and 193 nm double-exposure PR.

References

[1] Senefelder A 1911 History of stone-printing *The Invention of Lithography* (Whitefish, MT: Kessinger Publishing) section I

[2] Moore G E 2006 Progress in digital integrated electronics *IEEE Solid-State Circuits Soc. Newslett.* **11** 11–3
[3] Rothschild M 2005 Projection optical lithography *Materialstoday* **8** 18–24
[4] Okoroanyanwu U 2010 The semiconductor lithographic processChemistry and Lithography (Bellingham, WA: SPIE) ch 11
[5] Spiller E 1994 Imaging systems for x-raysSoft X-ray Optics (Bellingham, WA: SPIE) ch 4
[6] Levinson H J 2010 Photoresists *Principles of Lithography* 3rd edn (Bellingham, WA: SPIE) ch 3
[7] Raut H K, Venkatesan A G, Nair S and Ramakrishna S 2011 Anti-reflective coatings: a critical, in-depth review *Energy Environ. Sci.* **4** 3779–840
[8] Couteau T and Carcasi M 2006 Topside anti-reflective coating process and productivity improvements on KrF lithograph *Proc. SPIE* **6153** 61533H
[9] Rayleigh L F R S 1879 XXXI. Investigations in optics, with special reference to the spectroscope *The London, Edinburgh, and Dublin Philosophical Magazine and Journal of Science Series.* 5 **8** 261–74
[10] Kapoor R and Adner R 2007 Technology interdependence and the evolution of semiconductor lithography *Solid State Technol.* **50** 51–4
[11] Owa S, Nakano K, Nagasaka H, Fujiwara T, Matsuyama T, Ohmura Y and Magoo H 2007 Immersion lithography ready for 45 nm manufacturing and beyond 2007 IEEE/SEMI Advanced Semiconductor Manufacturing Conf. *(Stresa, Italy)* June, 2007 pp 238–44
[12] Mack C A 2005 Exploring the capabilities of immersion lithography *2005 IEEE/SEMI Advanced Semiconductor Manufacturing Conf. (Munich)* April, 2005 pp 1–6
[13] Pan D Z, Liebmann L, Yu B, Xu X and Lin Y 2015 Pushing multiple patterning in sub-10nm: are we ready? *2015 52nd ACM/EDAC/IEEE Design Automation Conf. (DAC) (San Francisco, CA)* June, 2015 pp 1–6
[14] Pikus F G and Torres A 2016 Advanced multi-patterning and hybrid lithography techniques *2016 21st Asia and South Pacific Design Automation Conf. (ASP-DAC) (Macau)* Jan., 2016 pp 611–6
[15] Raghunathan A, Fenger G, Lam M, Clifford C, Adam K and Sturtevant J 2017 Edge placement errors in EUV from aberration variation *Proc. SPIE* **10451** 104510W
[16] Mack C A 2009 Impact of mask roughness on wafer line-edge roughness *Proc. SPIE* **7488** 748828
[17] Postnikov S, Hector S, Garza C, Peters R and Ivin V 2003 Critical dimension control in optical lithography *Microelectron. Eng.* **69** 452–58
[18] Liu C H and Chen C L 2007 CMOS fabrications *Semiconductor Device Physics and Process: Theory & Practice* 2nd edn (Taipei: Wu-Nan Book Inc.) ch 6
[19] Levinson H J 2010 Extreme ultraviolet lithography *Principles of Lithography* 3rd edn (Bellingham, WA: SPIE) ch 12
[20] Masnovi M and Parchamy H 2019 Calculation of the extreme-ultraviolet radiation conversion efficiency for a laser-produced tin plasma source *Phys. Open* **1** 1–8
[21] Fomenkov I 2017 EUV source for high volume manufacturing: performance at 250W and key technologies for power scaling *2017 Source Workshop* (Dublin) Nov., 2017
[22] Mizoguchi H *et al* 2018 High power LPP-EUV source with long collector mirror lifetime for high volume semiconductor manufacturing *2018 China Semiconductor Technology Int. Conf. (CSTIC) (Shanghai)* Mar., 2018 pp 1–7

[23] Putna E S, Younkin T R, Chandhok M and Frasure K 2009 EUV lithography for 30 nm half pitch and beyond: exploring resolution, sensitivity and LWR tradeoffs *Proc. SPIE* **7273** 72731L

[24] Chawon K, Liping R, Jacque G, Frank G, Stefan W, Bill P, Joo-On P, Tom W, Todd R Y and Patrick N 2009 Assessment of EUV resist readiness for 32 nm hp manufacturing and extendibility study of EUV ADT using state-of-the-art resist *Proc. SPIE* **7271** 727124

[25] Itani T 2009 Recent status and future direction of EUV resist technology *Microelectron. Eng.* **86** 207–12

[26] Noga D E, Lawson R A, Lee C T, Tolbert L M and Henderson C L 2009 Understanding pattern collapse in high-resolution lithography: impact of feature width on critical stress *Proc. SPIE* **7273** 727334

[27] Cao H B, Nealey P F and Domke W D 2000 Comparison of resist collapse properties for deep ultraviolet and 193 nm resist platforms *J. Vac. Sci. Technol.* B **18** 3303–7

[28] Chandhok M, Frasure K, Putna E S, Younkin T R, Rachmady W, Shah U and Yueh W 2008 Improvement in linewidth roughness by postprocessing *J. Vac. Sci. Technol.* B **26** 2265–70

Chapter 2

Multilayer mirrors in extreme ultraviolet lithography

The diffraction limit depends on the light wavelength, so a shorter wavelength used in optical lithography can achieve a finer line width. The extreme ultraviolet (EUV) wavelength at 13.5 nm is therefore proposed as a central wavelength (CW) to be used in next-generation lithography (NGL) tools after immersion lithography. Since most material has strong absorption in this region, transparent optical lenses and photomasks are no longer applicable. Instead, multilayer (ML) structures are used as mirrors and as a substrate of reflective photomask in a EUV lithography system. In the large sense, the term 'mirror' in a EUV lithography system refers to both reflective mirror and photomask. This chapter will start with an introduction to the composition of the generic EUV mirror and photomask. The fabrication process for both of them will also be mentioned in the same section. Next, different types of EUV photomasks will be introduced in section 2.2. Then, how the ML mirror reflects EUV will be discussed in section 2.3. In section 2.4, the quarter-wave EUV mirror, which is a common mirror design, will be presented. In the last section, a design from a photonic crystal perspective will be investigated.

2.1 Composition and fabrication

Due to the high EUV absorption rate of almost all matter including air, the optics in an EUV lithography system must be built and operated in vacuum (wavelengths below 100 nm cannot be transmitted through air). Similarly, EUV radiation has extremely high absorption in almost all materials, so optical components in EUV applications including lenses and the photomask must be designed to be reflective. Therefore, the ML structure based mirrors with proper selected materials that can reflect EUV were developed. Different types of ML structure based mirrors were proposed in several published research in order to intensify the EUV reflectivity, and we will review the research later in section 2.3.1. Regardless, this kind of reflective

doi:10.1088/978-0-7503-2652-0ch2

optics makes both the fabrication process and materials different from optical components in between UV and visible regions. In a current deep ultraviolet (DUV) lithography system, patterns on the photomask are transferred onto the wafer through the projection optics with a demagnification factor of, typically, 4. Due to this demagnification, the constraints on the precision of photomask patterns was relaxed. For instance, a critical dimension (CD) of 40 nm on the photomask can be used to print 10 nm CD on the wafer. As the targeted CD of the integrated circuit (IC) chip shrinks in EUV lithography, so does the CD on the EUV photomask, making photomask fabrication more challenging. The fabrication process of both the EUV mirror and photomask can be seen in figure 2.1. In principle, the EUV mirror and photomask share the same reflective ML blank. The process to produce a ML mirror for EUV, as illustrated by red arrows, is indicated in figure 2.1(a)–(d). Furthermore, the process to produce an EUV photomask, as illustrated by blue arrows, can be seen in figure 2.1(a)–(h). Based on the reflective ML mirror, the EUV photomask can be further fabricated by adding an absorber layer and a buffer layer. The absorber layer is used to create photomask patterns, while the buffer layer serves as etch stop during the absorber layer etch. The below contexts are descriptions of each layer:

– **A backside conductive layer**: Significantly, an extremely careful handling and chucking of EUV mask or mirror is necessary in order to avoid mechanical abrasion and the formation of particles, which may deteriorate the

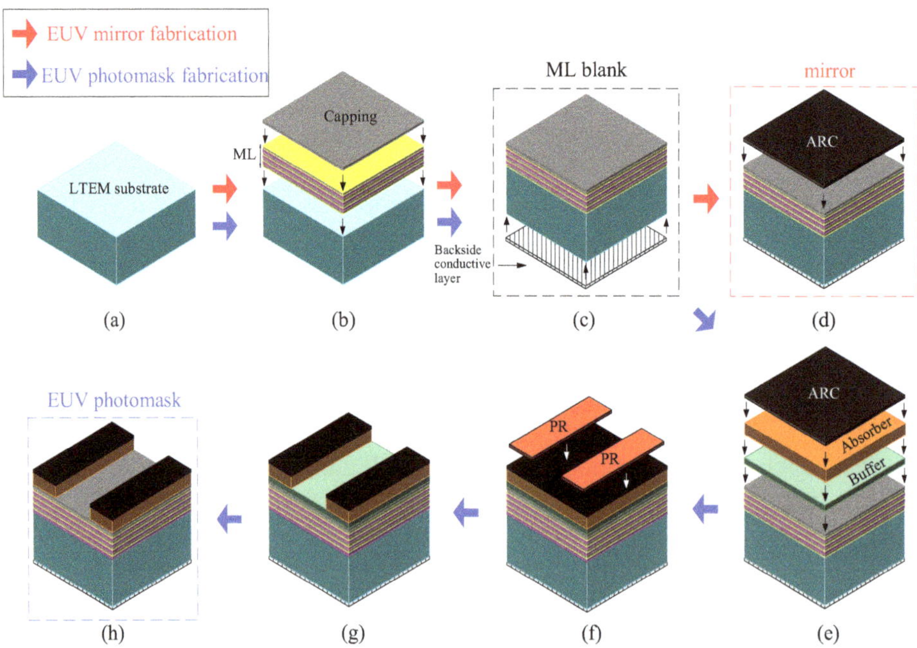

Figure 2.1. EUV mirror and photomask fabrication process. As indicated by the red arrows, (a)–(d) depict the EUV mirror fabrication process. As indicated by the blue arrows, (a)–(h) depict the EUV photomask fabrication process.

performance of an EUV lithography system. In order to accomplish these handling requirements, the EUV mask or mirror will be held by an electrostatic chuck in the system. As the substrate (or blank) of the EUV mask or mirror is a dielectric (usually glass), an electrically conductive layer has to be coated on the backside to be able to electrostatically clamp the substrate, as sketched in figure 2.1(c). In addition, to allow image placement correction by ultra-short pulsed laser, the backside coating has to be optically transparent and electrically conductive at the same time.

- **A substrate**: The substrate is also a dielectric, as seen from figure 2.1(a), it decides the size of the mirror and photomask. However, the production requirements of EUV substrate and resulting photomask or mirror need to support thermal loadings at up to ~6 watts for high system throughputs. There will be a high risk of uneven substrate due to thermal deformation. This results in image distortion of EUV lithography. Thus, the so-called low thermal expansion material (LTEM) is used (a quartz with a certain percent of TiO_2).
- **A ML structure**: The ML interference coatings, as shown in figure 2.1(b), on the top of a LTEM substrate are necessary to make the mirror/photomask reflective at EUV wavelength. At this moment, the ML is formed by depositing alternating layers of two materials that have different index of refraction. The design is to position strongly absorbing materials (i.e. absorber) in very thin layers into the nodes of the low intensity standing-wave field and fill the remaining high intensity field spaces with a low absorption material (i.e. spacer) [1–3].The physics to reflect EUV will be introduced in section 2.3.
- **A capping layer**: The ML mirrors are especially sensitive to physical and chemical damage, which can significantly reduce their reflectivity and optical quality. Reflectivity in the EUV region is already low as compared to other reflectors for longer wavelengths, which is a particular problem since a typical EUV lithography system has several mirrors plus the reflecting photomask as indicated in section 1.4 of chapter 1. It is therefore evident that even a small decrease in the peak reflectivity of a single mirror will cause a significant light throughput reduction. A further problem is that plasma based EUV sources are typically dirty as they also emit a significant amount of fast ions and other particles. These can damage ML mirrors in the illumination system. To minimize the above issues, a capping layer as shown in figure 2.1(b) is therefore coated above to resist chemical and physical attacks.
- **A buffer layer**: As shown in figure 2.1(e), a buffer layer is used to protect the ML during the absorber layer etch and repair. It is not necessary to use a buffer layer if both absorber etch and repair have a sufficient selectivity to the ML capping. A buffer layer should have no pinhole defect and relatively low EUV absorption such that a large buffer defect can be tolerated without repair. Besides, it needs to be deposited with low stress to minimize stress-induced mirror or photomask, which induces a non-flatness mirror and this non-flatness will translate to the pattern in wafer printing.

- **An absorber layer**: A typical EUV photomask absorber layer stack consists of a metal, which absorbs the EUV light during exposure. The EUV light reflection at the absorber is typically less than 0.2%, it is deposited on top of the photomask for patterning purposes as shown in figure 2.1(e).
- **An anti-reflective coating (ARC) layer**: It can be seen in both figures 2.1(d) and (e) that the antireflection function of the ARC layer is to reduce the light reflection from the top of the photomask and mirror defect inspection wavelength, typically the DUV wavelength will be used. In the case of a reflective EUV photomask, the inspection image contrast depends on the reflected light between the ML region and the absorber region. To incorporate the inspection need, a low DUV reflectivity ARC is needed.

2.1.1 ML mirror fabrication

The material selections of ML mirror are based on the semiconductor equipment and material international (SEMI) standard [4]. As for the fabrication process, figures 2.1(a)–(d) show the fabrication process of a ML mirror, while figures 2.1(e)–(h) show the process of photomask patterning. Light scatter is higher at short wavelengths, which make surface roughness, surface unflatness, and other surface tolerances important for EUV optics. Thus, the fabrication process of a ML mirror starts with a full polish of the LTEM. A well-polished LTEM not only enhances the surface flatness of the LTEM substrate as well as reduces the possibility of particle attachment but also prevents stray lights. According to SEMI P37-1102, the LTEM substrate is made of Zerodur with the allowed thermal expansion coefficient (TEC) in between ±30 ppb K^{-1}. The LTEM deformation caused by EUV radiation is about to happen if TEC of the selected LTEM substrate is out of the allowed values. This LTEM deformation will result in shape distortions and affect image quality at the wafer level. They are basically characterized from two types:

- **In-Plane distortion (IPD)**: Besides EUV radiation, this can also be induced by differences in the chucking mechanisms during transportation to the scanner. As shown in figure 2.2(a), bending results in an IPD along the *x*-direction of

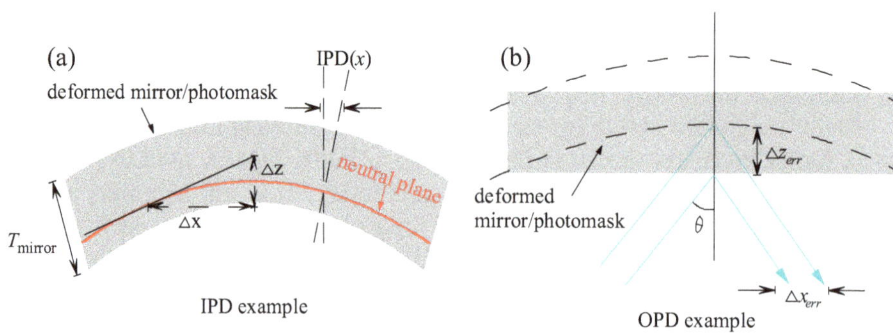

Figure 2.2. Two types of LTEM deformation: (a) in-plane distortion (IPD) in the *x*-direction, (b) out-of-plane distortion (IPD) in the *x*-direction.

the mirror/photomask surface depends on the local slope $\frac{\Delta z}{\Delta x}$ or $\frac{\Delta z}{\Delta y}$. The IPD can be described as

$$\mathrm{IPD}(x, y) = \alpha \cdot T_{\mathrm{mirror}} \cdot \frac{\Delta z}{\Delta(x, y)}, \qquad (2.1)$$

where α stands for the factor of neutral plane. Δx, Δy and Δz stand for the deformation along the x-, y- and z-directions. T_{mirror} is the thickness of the mirror/photomask.

– **Out-of-Plane distortion (OPD)**: As shown in figure 2.2(b), assuming the EUV light illuminates a ML mirror with an angle of θ, the image error Δx_{err} (assume in x-direction) at the wafer level caused due to LTEM substrate deformation in z-direction can be described as [5, 6]

$$\Delta x_{err} = \frac{\Delta z_{err} \cdot \tan\theta}{x_r}. \qquad (2.2)$$

where x_r stands for the image reduction ratio of the EUV lithography system and Δz_{err} stands for the LTEM substrate deformation in the z-direction. The Δx_{err} can also be induced by the imperfect LTEM polish, which is so-called LTEM surface roughness. The issues of LTEM surface roughness include not only Δx_{err} but also the scattering angle of the EUV, the latter decreases the reflectivity of the EUV ML mirror which is a critical performance loss. According to SEMI P37-1102, the roughness is defined as peak-to-valley flatness (PVF). The allowed PVF value on a LTEM substrate with a size of 142 mm^2 is ~50 nm [7].

After preparing the LTEM substrate, alternative thin films with relative low and high refractive indexes are repeatedly deposited above the well-prepared LTEM substrate. Currently, molybdenum (Mo, low index) and silicon (Si, high index) paired with 40 periods is the most applied EUV ML structure. This ML structure is the key component to reflect EUV. After ML deposition, ruthenium (Ru) or Si, which has low absorption to EUV, will be deposited as a capping layer to minimize the EUV absorption of the ML mirror. Then, a backside conductive layer will be coated at the LTEM substrate bottom to enhance the efficiency of the electrostatic chuck. Before completing ML mirror fabrication, an ARC layer which has low reflection to DUV is typically coated if DUV inspection will be applied.

2.1.2 Photomask fabrication

The photomask in the EUV lithography system is a reflective optical component, the fabrication process is based on a ML mirror. In order to prevent the potential ML damages from upcoming patterning process, the buffer layer is coated above the capping layer. From numerous research, the buffer layer can be SiO_2, SiON, carbon (C), chromium (Cr) and Ru [6, 8–15]. More attention needs to be paid to the buffer layer coating process, because it is easily to induce unfavorable stresses which could twist the

ML mirror which results in photomask IDP. According to the standard of SEMI P38-1103, the stress should be controlled ⩽200 MPa. Afterward, the absorber layer is coated for patterning purposes. The material of the absorber layer can be AlCu, titanium (Ti), TiN, tantalum (Ta), TaN and Cr [6, 8–11, 13, 16–20]. Regardless, another benefit of the absorber layer is to absorb massive EUV exposures. This function prevents the photomask from unsatisfied distortions due to overheating. Then, the same as the ML mirror, an ARC layer that has low reflection to DUV is coated if DUV inspection is applied. The last process of EUV photomask fabrication is the photomask patterning process, which can be classified into three steps:

- **Photomask patterning through electron writing:** This process is just like the fabrication process of transparent photomasks. This process is so-called electron-beam lithography where a computer-aided design (CAD) is transferred to a PR layer. An electron gun uses controlled electrons together with the high precision stage to strike resists to form a pattern directly according to CAD demands. As electron guns emit a large number of electrons, the pathways of these electrons within are controlled by electron lenses. Generally speaking, electron lenses are metallic structures with circular shapes that allow electrons to pass through their centers. As the desired electric voltages are applied on electron lenses, electrons can be controlled and focused on the designed focal plane. So that electrons are able to precisely form a focused spot and transfer electrons onto the PR, as illustrated in step 1 in figure 2.3. Owing to the higher resolution requirement compared to transparent photomasks, the voltage used for the electron-beam writing system must be higher than 50 keV. In addition, as seen in step 2 in figure 2.3, the temperature of the post exposure bake has to be well-controlled and below 150 °C.
- **Absorber layer etching:** As shown in step 4 in figure 2.3, the dry etching is used to etch the absorber layer on the EUV photomask. Two different dry etching processes exist depending on the material selection of absorber and ARC layer. If Cr is selected as an absorber and $C_xO_yN_z$ is selected as an ARC, the dry etching process is almost the same as etching a transparent photomasks, which is a chemical methodology based on chlorine (Cl_2) gas and oxygen (O_2) [5, 16]. If TaN is chosen as an absorber, the dry etching is based on Cl_2 gas [5, 20].
- **Buffer layer etching:** This process for EUV photomask is a bit different from a transparent photomask. As illustrated in step 5 in figure 2.3, additional PR film as a hard mask is not required. Instead, the absorber layer that is patterned by CAD layouts is used as a hard mask during buffer layer etching. Different dry etching techniques are applied according to material selections: Cl_2 or fluorine (F_2) gas can be used in chemical etching if the buffer layer is selected as SiO_2 [5, 21]. If Cr or Ru is selected as a buffer layer, Cl_2 gas is combined with O_2 in chemical etching [5]. If C is selected as a buffer layer, the O_2 plasma needs to be used for buffer layer etching.

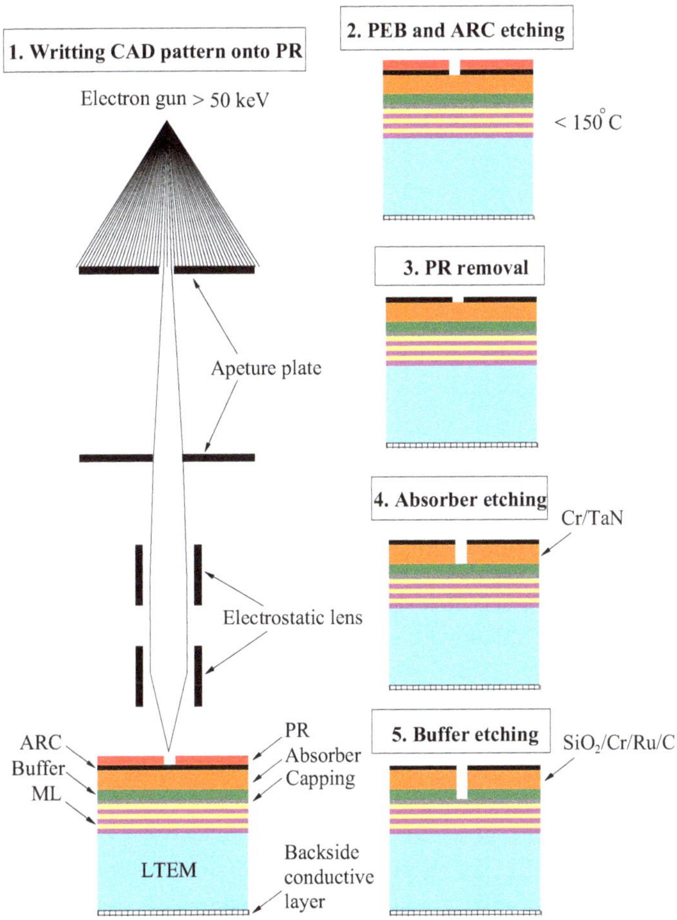

Figure 2.3. Generic EUV photomask patterning process, which combines e-beam writing, ARC layer etching, absorber layer etching and buffer layer etching.

2.2 ML based EUV photomasks

A reflective EUV photomask or reticle is a reflective version of the photomask. On a transmission photomask, as mentioned in section 1.1, the light is allowed to travel through the quartz substrate, and on the other hand the light is blocked by means of absorber patterns. On the EUV photomask, the light is reflected by the ML and reflected by absorber patterns. There are three types of EUV photomasks:

- **Binary intensity photomask (BIM)**: A binary photomask is the earliest concept of photomask and it is also very common in a 193 nm immersion lithography system. As shown in figure 2.4(a), a BIM for EUV is composed of equal thickness absorber features on the ML mirror. Light reflects from the ML areas and is absorbed by the absorber areas. Please note that EUV typically illuminates the photomask at 6° in order to produce the peak reflectivity. Thus, there will be a 6° shift in the resulting electric fields after BIM, electric

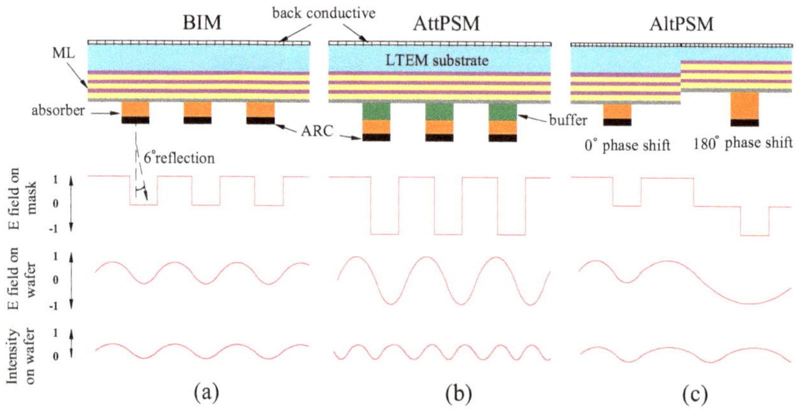

Figure 2.4. Theoretical electric field and intensity from different types of ML mirror based EUV photomasks, the 6° EUV illumination is applied to all of them. (a) Binary intensity mask, (b) attenuated phase-shift mask and (c) alternating phase-shift mask.

field on wafer and the intensity on wafer. In this example, there is significant energy drops below the absorber areas. However, due to the close proximity effect between the neighboring ML and absorber areas, the contrast of the electric field on the wafer becomes lower than the intensity. This unwanted energy effects the quality of intensity on wafer, and thus affects resist profiles. Therefore phase-shift techniques are designed to compensate this drawback.

– **Attenuated phase-shift photomask (AttPSM)**: An EUV AttPSM increases the intensity contrast in the electric field by means of thickening absorber layer. Cr was used as an absorber of a transmission AttPSM. Owing to high absorption in the EUV region, it is difficult to cherry-pick a material as an absorber of EUV AttPSM that also meets both the required intensity attenuation and phase shift. Therefore, two different materials are typically used to compose a two-layer absorber for EUV AttPSM, as sketched in figure 2.4(b). When EUV propagates towards two absorber layers, the intensity attenuations can be described as [5]:

$$I_r = I_i \left| \frac{-2\pi}{\lambda} \left(\frac{2k_{a1}T_{a1}}{\cos\theta} + \frac{2k_{a2}T_{a2}}{\cos\theta} \right) \right|^2, \tag{2.3}$$

where I_i stands for the intensity of EUV illumination, I_r stands for the resulting EUV intensity reflected from the top absorber layer, λ stands for EUV wavelength, k_{a1} and k_{a2} stand for the extinction coefficient (or attenuation coefficient) of two absorber layers, T_{a1} and T_{a2} stand for the thickness of two absorber layers and θ is the illumination angle. Based on equation (2.3), the intensity attenuation of EUV AttPSM, which can be described as $\frac{I_r}{I_i}$, is determined by not only the absorption rate of two absorber layers but also their thickness. Nevertheless, the absorber layer above ML blank of the EUV AttPSM are sometimes replaced by a buffer layer for different applications. The phase shift can be described as (assuming phase shift = 180°)

$$\frac{2\pi}{\lambda}\left(\frac{2T_{a1}(1-n_{a1})}{\cos\theta}+\frac{2T_{a2}(1-n_{a2})}{\cos\theta}\right)=\pi. \tag{2.4}$$

In equation (2.4), n_{a1} and n_{a2} stand for refractive index of two absorber layers. By utilizing equations (2.3) and (2.4) with required phase shift, either the thickness or the materials of two absorber layers of an AttPSM can be designed. The resulted dense intensity on the wafer can be seen in figure 2.4(b). Owing to the thicker absorber, the electric fields of the absorber areas fall below 0. Therefore, the peaks of the intensity on the wafer can be more compared to the BIM, and thus increases the resolution of the resist profiles.

- **Alternating phase-shift photomask** (AltPSM): For a transmission AttPSM which uses transparent fused silica as its substrate blank, the 180° phase-shift effect is generated by etching the absorber layer during the process of photomask pattern definition. For an EUV AttPSM, however, the way to produce 180° phase-shift is different. To produce 180° phase-shift for an EUV AttPSM, the 180° phase step needs to be created on the LTEM substrate prior to the ML fabrication process as shown in figure 2.4(c). There are two ways to produce a phase step before ML deposition, one is to etch LTEM directly, the other is to etch the film that was pre-deposited above the LTEM substrate. Equation (2.5) defines the required etching depth of the designed phase-shift [5, 22]:

$$T_m=\frac{(2m+1)\lambda}{4\cos\theta}, \tag{2.5}$$

where θ stands for the illumination angle of EUV. The value of m can be tuned according to the required specification, it can be one of any positive integer (m = 0, 1, 2 ...). For instance, if the maximum allowed thickness difference on the surface of EUV AttPSM is 3.5 nm. Then, the designer can use $m = 0$ to design the AttPSM because by substituting $m = 0$, equation (2.5) gives T_m = 3.36 nm. Apparently, the smaller value of m yields the thinner phase step, but the tradeoff is that it is more likely to induce surface unflatness. The intensity profile in figure 2.4(c) describes how AttPSM provided works in different way, the 180° phase shift enables negative electric field on photomask and wafer, which induces a phase lag in intensity on the wafer. Therefore, provides flexibility to distribute intensity peaks on the wafer.

2.3 Multilayer mirror physics

The ML mirror is composed of multiple thin layers of dielectric material. Therefore, the ML mirror is also called the dielectric ML mirror or just dielectric ML. A ML stack becomes a type of mirror in a selected wavelength region when it utilizes multiple interferences of low and high material indices to achieve the desired reflectivity, and the high reflectivity contributes from constructive interferences. Among all the composed layers, the ML contributes to the reflectivity of a ML mirror. Owing to the high attenuating of EUV especially at the desired wavelength of 13.5 nm for EUV lithography, only ML structures are found to reflect the energy throughout the exposure system. Thus far, reflectivity limited at approximately 70% can be achieved with small illuminated angle (i.e. about 6°) at $\lambda = 13.5$ nm by using an approximately 40 periodic Mo/Si ML based mirror [5, 23–26], although from our literature reviews the Mo/Si based EUV ML mirror is not the only type. In some research, a few subsidiary layers added in the Mo/Si based ML mirror were also reported [27, 28].

2.3.1 Literature review of high-reflectivity EUV mirror

Up until now, the search for different materials and combinations of the EUV ML mirror to result in higher reflectivity is still a continuous process, because low conversion efficiency and high absorption rate of the EUV light reflect a heavy cost in EUV lithography. Only when the EUV light source power can achieve the demanded level [2, 29], can the cost of EUV lithography under HVM become much more competitive than that of other lithography tools [30]. Nevertheless, it is not hard to estimate that the EUV power can loss significantly after several reflections in such multi-mirror system. Here are examples: Carniglia *et al* mentioned that it is possible to exceed the reflectance limit of the quarter-wavelength mirror design if we alter the layer thickness of the constructed materials [31]. Singh *et al* optimized the reflectivity of the Ru–Mo/strontium (Sr)–rubidium (Rb) based ML mirror [32]. They concluded that if a ML has more composed materials, it has the potential to reach the better performances. However, the fabrication becomes much more complicated, costly, and time-consuming when there are more than two composed materials. A similar case was proposed by Kang *et al* when they optimized the reflectivity of the EUV mirrors [33]. Modi *et al* presented that by using a NbC/Si ML mirror, where the stable reflectivity can be preserved, and the highest reflectivity from their results is 70.6% [34]. Maury *et al* indicated that the change of the Mo thickness turns out to be the transition between the amorphous to crystalline state of the Mo layers. This transition results in the variation of the refractive index, thus it affects the reflectivity of the Mo/Si ML mirror [35]. Thus, the reflectivity can be increased by adding other materials or adjusting the thickness of the EUV MLs. Other researchers studied the aperiodic Mo/Si ML mirrors [36, 37].

2.3.2 EUV light interacts with ML mirror

For reflective properties to work, the ML mirrors obey Fresnel's reflection and transmission principles. Since the periodic Mo/Si ML based mirror is now the main stream in EUV lithography, we use it to describe the physics. As shown in

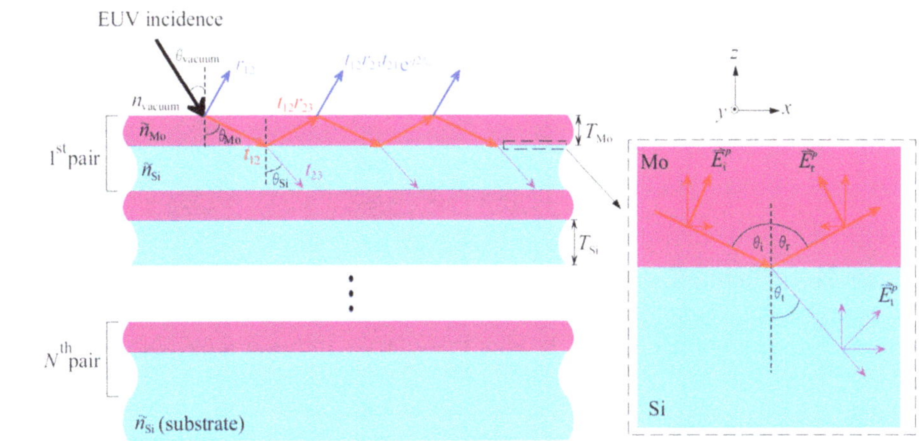

(a) (b)

Figure 2.5. One-dimensional Mo/Si ML stack with N pairs that is deposited on a Si substrate, x- and y-directions are assumed to be infinitely long. In which (a) EUV light reflects and refracts from the first two interfaces when it interacts with a Mo/Si based ML blank, several of the partial waves are shown. (b) The electric field vectors interact between the Mo/Si boundary. Only the components in the plane of incidence (x–z) are shown, the component perpendicular to the plane of incidence (y) is not shown.

figures 2.5(a) and (b), assuming we have a one-dimensional Mo/Si ML stack with N pairs that is deposited on a Si substrate along the z-direction. Please note that both z- and y-directions are assumed to be infinitely long. Now imagine a EUV light propagates in the vacuum and eventually illuminates the Mo film on top of the ML stack at a specific grazing angle θ_{vacuum}. To see how EUV reflection works in terms of electromagnetic fields, we shall firstly look into the interactions of electric and magnetic fields of the topmost Mo/Si interface as shown in figure 2.5(b). Taking the plane of incidence as the x–z-plane with the y-direction going out of the page. We separate the electric field vector into components in the plane of incidence, which is the so-called s-wave and also referred to as the transverse-electric (TE) wave. In figure 2.5(b), we show the electric vectors in the plane of incidence, which is the p-wave and also referred to as the transverse-magnetic (TM) wave. The amplitude of the incident wave is denoted as $|\vec{E}_{\text{i}}^{\,p}|$, the amplitude of the reflected wave is denoted as $|\vec{E}_{\text{r}}^{\,p}|$, and the amplitude of the transmitted wave (or refractive) is denoted as $|\vec{E}_{\text{t}}^{\,p}|$. Please note that $|\vec{E}_{\text{i}}^{\,s}|$, $|\vec{E}_{\text{r}}^{\,s}|$ and $|\vec{E}_{\text{t}}^{\,s}|$ are the corresponding amplitudes perpendicular to the plane of incidence, which are not shown in the plot. The components which are parallel to the surface in the x-direction are denoted as $\vec{E}_{\text{i}}^{\,x}$, $\vec{E}_{\text{r}}^{\,x}$ and $\vec{E}_{\text{t}}^{\,x}$. The corresponding quantities for the magnetic wave are denoted as $\vec{H}_{\text{i}}^{\,p}$, $\vec{H}_{\text{r}}^{\,p}$, $\vec{H}_{\text{t}}^{\,p}$, $\vec{H}_{\text{i}}^{\,s}$, $\vec{H}_{\text{r}}^{\,s}$, $\vec{H}_{\text{t}}^{\,s}$, $\vec{H}_{\text{i}}^{\,x}$, $\vec{H}_{\text{r}}^{\,x}$ and $\vec{H}_{\text{t}}^{\,x}$. By definition, the Fresnel reflection coefficients of p- and s-polarizations are

$$r^p = \frac{|\vec{E}_{\mathrm{r}}^{\,p}|}{|\vec{E}_{\mathrm{i}}^{\,p}|} \text{ and } r^s = \frac{|\vec{E}_{\mathrm{r}}^{\,s}|}{|\vec{E}_{\mathrm{i}}^{\,s}|}. \tag{2.6}$$

The Fresnel transmission coefficients of p- and s- polarizations are

$$t^p = \frac{|\vec{E}_{\mathrm{t}}^{\,p}|}{|\vec{E}_{\mathrm{i}}^{\,p}|} \text{ and } t^s = \frac{|\vec{E}_{\mathrm{t}}^{\,s}|}{|\vec{E}_{\mathrm{i}}^{\,s}|}. \tag{2.7}$$

The boundary condition that must be satisfied is that the sum of the tangential components of the electric fields just above the interface must equal the sum of the components just below the interface. This boundary condition must also be applied to the magnetic field components. In equation form, these boundary conditions are described as

$$|\vec{E}_i^{\,x}| + |\vec{E}_r^{\,x}| = |\vec{E}_t^{\,x}|, \tag{2.8}$$

$$|\vec{E}_i^{\,y}| + |\vec{E}_r^{\,y}| = |\vec{E}_t^{\,y}|, \tag{2.9}$$

$$|\vec{H}_i^{\,x}| + |\vec{H}_r^{\,x}| = |\vec{H}_t^{\,x}| \tag{2.10}$$

and

$$|\vec{H}_i^{\,y}| + |\vec{H}_r^{\,y}| = |\vec{H}_t^{\,y}|. \tag{2.11}$$

With substitution, the left-hand side of equation (2.8) becomes

$$|\vec{E}_i^{\,p}| \cos\theta_i + |\vec{E}_r^{\,p}| \cos\theta_r = \left(|\vec{E}_i^{\,p}| - |\vec{E}_r^{\,p}|\right)\cos\theta_i, \tag{2.12}$$

where θ_{i} and θ_{r} stand for the angle of incidence and reflection of the EUV light in the Mo layer. Since $\cos\theta_r = -\cos\theta_i$. There is only one tangential component below the interface, that is, $|\vec{E}_t^{\,p}|\cos\theta_t$, and hence equation (2.8) becomes

$$\left(|\vec{E}_i^{\,p}| - |\vec{E}_r^{\,p}|\right)\cos\theta_i = |\vec{E}_t^{\,p}| \cos\theta_t, \tag{2.13}$$

where θ_t stands for the angle of refraction of the EUV light in the Si layer. For the s-wave, equation (2.9) gives us simply

$$|\vec{E}_i^{\,s}| + |\vec{E}_r^{\,s}| = |\vec{E}_t^{\,s}|. \tag{2.14}$$

Similarly, for the magnetic components, equation (2.10) becomes

$$\left(|\vec{H}_i^{\,p}| - |\vec{H}_r^{\,p}|\right)\cos\theta_i = |\vec{H}_t^{\,p}| \cos\theta_t, \tag{2.15}$$

and equation (2.11) becomes

$$|\vec{H}_i^{\,s}| + |\vec{H}_r^{\,s}| = |\vec{H}_t^{\,s}|. \tag{2.16}$$

The magnetic field is not independent of the electric field. In general, in vector format, $\vec{H} = \tilde{n}\vec{s} \times \vec{E}$, where $\vec{s}$ is the unit vector of the direction of propagation and $\tilde{n}$

is the complex index of refraction. Substituting this into equations (2.15) and (2.16) yields

$$\tilde{n}_{\mathrm{Mo}}\left(|\vec{E}_{\mathrm{i}}^{s}| - |\vec{E}_{\mathrm{r}}^{s}|\right)\cos\theta_{\mathrm{i}} = \tilde{n}_{\mathrm{Si}}\,|\vec{E}_{\mathrm{t}}^{s}|\cos\theta_{\mathrm{t}} \tag{2.17}$$

and

$$\tilde{n}_{\mathrm{Mo}}\left(|\vec{E}_{\mathrm{i}}^{p}| + |\vec{E}_{\mathrm{r}}^{p}|\right) = \tilde{n}_{\mathrm{Si}}\,|\vec{E}_{\mathrm{t}}^{p}|, \tag{2.18}$$

where $\tilde{n}_{\mathrm{Mo}}$ and $\tilde{n}_{\mathrm{Si}}$ stand for complex index of refraction of Mo and Si. Now equations (2.13) and (2.14) stand for boundary conditions dealing with components parallel to the plane of incidence, and equations (2.17) and (2.18) stand for boundary conditions dealing with components perpendicular to the plane of incidence. By applying the definition of Fresnel reflection and transmission, equations (2.13), (2.14), (2.17) and (2.18) can be solved to give

$$t^{p} = \frac{2\tilde{n}_{\mathrm{Mo}}\cos\theta_i}{\tilde{n}_{\mathrm{Si}}\cos\theta_i + \tilde{n}_{\mathrm{Mo}}\cos\theta_t}, \tag{2.19}$$

$$t^{s} = \frac{2\tilde{n}_{\mathrm{Mo}}\cos\theta_i}{\tilde{n}_{\mathrm{Mo}}\cos\theta_i + \tilde{n}_{\mathrm{Si}}\cos\theta_t}, \tag{2.20}$$

$$r^{p} = \frac{\tilde{n}_{\mathrm{Si}}\cos\theta_i - \tilde{n}_{\mathrm{Mo}}\cos\theta_t}{\tilde{n}_{\mathrm{Si}}\cos\theta_i + \tilde{n}_{\mathrm{Mo}}\cos\theta_t}, \tag{2.21}$$

and

$$r^{s} = \frac{\tilde{n}_{\mathrm{Mo}}\cos\theta_i - \tilde{n}_{\mathrm{Si}}\cos\theta_t}{\tilde{n}_{\mathrm{Mo}}\cos\theta_i + \tilde{n}_{\mathrm{Si}}\cos\theta_t}. \tag{2.22}$$

More generally, equations (2.19)–(2.22) stand for reflection and transmission coefficients for EUV light in Mo approaching an interface between Mo and Si. The ML mirror construction uses alternating pairs of selected materials to optimize the refractive index at the EUV wavelength of choice. As seen from figure 2.5(a), an EUV light is incident from a vacuum (black arrow) onto the topmost Mo film. For convenience, let us take the amplitude of the incident light to be unity. The amplitude of the first reflection will then be r_{12}, and the amplitude of the transmitted light in the topmost Mo layer after this first interaction will be t_{12}. If only one interface were present, the transmitted light would continue and never have to be dealt with again. However, when a second interface in present, some of this light are transmitted into the next Si layer and some are reflected back into the topmost Mo layer, heading back toward the upper interface. The amplitude of the returning partial wave is $t_{12}r_{23}$. When this beam interacts with the upper interface, some are transmitted into top Mo and some are reflected, heading again toward the lower interface. The amplitude of the part transmitted into the top Mo will then be $t_{12}r_{23}t_{21}$. This second partial wave will be out of phase with the first partial wave and the complex amplitude must include the phase factor $e^{-j2\beta_{\mathrm{Mo}}}$, where β_{Mo} is the phase change that the wave experiences as it traverses the top Mo film once between interfaces. The phase angle β_{Mo}, or the film phase thickness, is given by

$$\beta_{Mo} = 2\pi\left(\frac{T_{Mo}}{\lambda}\right)\tilde{n}_{Mo}\cos\theta_{Mo}, \quad (2.23)$$

where T_{Mo} is the thickness of Mo layers and θ_{Mo} stands for the angle of refraction of EUV light in Mo layers. The complex amplitudes of the successive partial waves are

$$r_{12}, \quad (2.24)$$

$$t_{12}r_{23}t_{21}e^{-j2\beta_{Mo}}, \quad (2.25)$$

$$t_{12}r_{23}r_{21}r_{23}r_{21}r_{23}t_{21}e^{-j4\beta_{Mo}}, \quad (2.26)$$

and so on.

Addition of the partial waves lead to an infinite geometric series for the total reflected amplitude. After rearranging some of the terms, we have the series

$$R_{top} = r_{12} + t_{12}r_{23}t_{21}e^{-j2\beta_{Mo}} + t_{12}r_{23}r_{21}r_{23}r_{21}r_{23}t_{21}e^{-j4\beta_{Mo}} + \dots, \quad (2.27)$$

where R_{top} stands for total reflection coefficient of the topmost Mo surface, which can be expressed as

$$R_{top} = r_{12} + t_{12}t_{21}r_{23}e^{-j2\beta_{Mo}}[1 + (r_{21}r_{23}e^{-j2\beta})^1 + \dots]. \quad (2.28)$$

By using the relationship

$$\frac{1}{1-x} = 1 + x + x^2 + x^3 + \dots(x^2 < 1), \quad (2.29)$$

we can obtain the total reflection coefficient

$$R_{top} = \frac{r_{12} + r_{23}e^{-j2\beta_{Mo}}}{1 + r_{12}r_{23}e^{-j2\beta_{Mo}}}. \quad (2.30)$$

In a similar manner, we can obtain the total transmission coefficient of the topmost Mo surface, which is given by

$$T_{top} = \frac{t_{12}t_{23}e^{-j\beta_{Mo}}}{1 + r_{12}r_{23}e^{-j2\beta_{Mo}}}. \quad (2.31)$$

Equations (2.30) and (2.31) are the total reflection coefficient and transmission coefficient for a film-covered surface. To calculate the reflection and transmission for an aperiodic ML stack, the total reflection coefficient needs to be iteratively applied to each interface between layers. For periodic MLs consisting of two alternative materials, there is a pattern to follow. Based on equations (2.21)–(2.23), (2.30) and (2.31), for both *s*- and *p*- polarizations, we can have the below characteristic matrix corresponding to one Mo/Si period

$$M_{\text{Mo/Si}} = \begin{bmatrix} \cos\beta_{\text{Si}}\cos\beta_{\text{Mo}} - \frac{g_{\text{Mo}}}{g_{\text{Si}}}\sin\beta_{\text{Si}}\sin\beta_{\text{Mo}} & -\frac{i}{g_{\text{Si}}}\cos\beta_{\text{Mo}}\sin\beta_{\text{Si}} - i\frac{i}{g_{\text{Mo}}}\sin\beta_{\text{Mo}}\cos\beta_{\text{Si}} \\ -ig_{\text{Mo}}\sin\beta_{\text{Mo}}\cos\beta_{\text{Si}} - ig_{\text{Si}}\cos\beta_{\text{Mo}}\sin\beta_{\text{Si}} & \cos\beta_{\text{Si}}\cos\beta_{\text{Mo}} - \frac{g_{\text{Si}}}{g_{\text{Mo}}}\sin\beta_{\text{Si}}\sin\beta_{\text{Mo}} \end{bmatrix} \tag{2.32}$$

where $\beta_{\text{Si}} = 2\pi\left(\frac{T_{\text{Si}}}{\lambda}\right)\tilde{n}_{\text{Si}} \cos\theta_{\text{Si}}$, T_{Si} stands for the thickness of Si layers and θ_{Si} stands for the angle of refraction of EUV light in Si layers. For *s*-polarization, $g_{\text{Si}} = \tilde{n}_{\text{Si}} \cos\theta_{\text{Si}}$ and $g_{\text{Mo}} = \tilde{n}_{\text{Mo}} \cos\theta_{\text{Mo}}$. For *p*-polarization, $g_{\text{Si}} = \frac{\cos\theta_{\text{Si}}}{\tilde{n}_{\text{Si}}}$ and $g_{\text{Mo}} = \frac{\cos\theta_{\text{Mo}}}{\tilde{n}_{\text{Mo}}}$. In addition, the characteristic matrix can be extended to a mirror consisting of N pairs of Mo/Si

$$(M_{\text{Mo/Si}})^N = \begin{bmatrix} M_{11} & M_{12} \\ M_{21} & M_{22} \end{bmatrix}. \tag{2.33}$$

Then, the reflection and transmission coefficients of a periodic mirror consisting of N pairs of Mo/Si are

$$r_{\text{Mo/Si}}^N = \frac{(M_{11} + M_{12}g_{\text{sub}})g_{\text{vacuum}} - (M_{21} + M_{22}g_{\text{sub}})}{(M_{11} + M_{12}g_{\text{sub}})g_{\text{vacuum}} + (M_{21} + M_{22}g_{\text{sub}})} \tag{2.34}$$

and

$$t_{\text{Mo/Si}}^N = \frac{2g_{\text{vacuum}}}{(M_{11} + M_{12}g_{\text{sub}})g_{\text{vacuum}} + (M_{21} + M_{22}g_{\text{sub}})}. \tag{2.35}$$

For s-polarization, $g_{\text{vacuum}} = n_{\text{vacuum}} \cos\theta_{\text{vacuum}}$ and $g_{\text{sub}} = \tilde{n}_{\text{sub}} \cos\theta_{\text{sub}}$. For p-polarization, $g_{\text{vacuum}} = \frac{\cos\theta_{\text{vacuum}}}{\tilde{n}_{\text{vacuum}}}$ and $g_{\text{sub}} = \frac{\cos\theta_{\text{sub}}}{\tilde{n}_{\text{sub}}}$. Where n_{vacuum} and n_{sub} stand for the refractive index of the vacuum and the substrate, θ_{sub} stands for the angle of refraction of EUV light in the substrate. The reflectivity of entire periodic mirror consists of N pairs of Mo/Si is

$$R_{\text{total}} = \left(r_{\text{Mo/Si}}^N\right)^2. \tag{2.36}$$

2.3.3 Quick reflectivity assessment through one-dimensional analytical solution

The above derived reflection coefficient interprets the analytical solution of the one-dimensional (1D) ML structure (for instance, as shown in figure 2.5(a), infinite long in both *x*- and *y*-directions), no scattering effect will be considered. Therefore, the

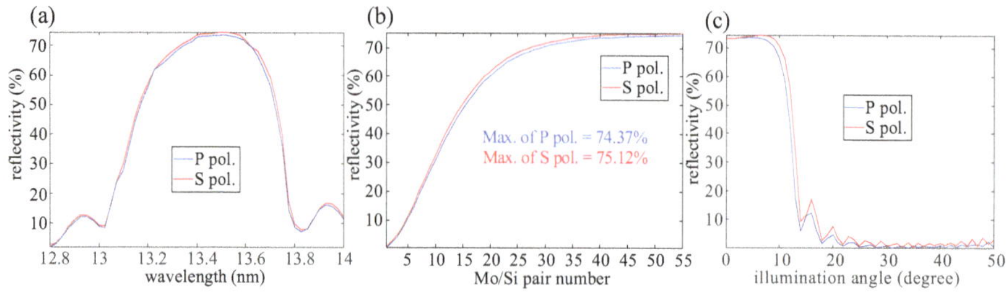

Figure 2.6. Calculated reflectivity based on one-dimensional analytical equation, in which (a) the reflectivity of a 40 Mo/Si ML based mirror under 6° illumination as a function of wavelength ranging from 12.8 nm to 14 nm, (b) the reflectivity of a Mo/Si ML based mirror under 6° illumination as a function of Mo/Si pair number and (c) the reflectivity of a 40 Mo/Si ML based mirror under different illuminations.

calculated reflectivity is a quick result from a simplified ML model. Anyhow, this analytical calculation is useful when we need a quick reflectivity assessment for a preliminary design concept. The methodology to perform rigorous simulations for ML will be introduced in chapter 3. By applying the above derived reflection coefficient, one can rapidly calculate the reflectivity of the entire Mo/Si ML. The calculations were done on a hexa core 2.60 GHz Intel i7-9750H CPU with 8 GB DDR4 RAM, and each calculation was done in a few milliseconds. Based on equation (2.36), the calculated reflectivity as a function of wavelength, Mo/Si pair and EUV illumination angle of an one-dimensional Mo/Si based ML mirror for both polarizations are calculated in figures 2.6(a)–(c). Table 2.1 provides the thickness information for the calculation. The thickness of Mo and Si can be designed through verities of methodologies [5]. Thus far, the main stream of the optimized thickness of Mo and Si for EUV lithography are 2.78 nm and 4.17 nm [6, 7, 9], we also used these thicknesses to calculate figure 2.6. One thing to mention here is that the thickness of the substrate is assigned to be infinite in calculations since it will always be far larger than the thickness of other layers (typically 600 μm–700 μm). The complex refractive index of Mo, Si, and backside coating Cr used to calculate figure 2.6 are shown in figure 2.7 [38]. The reflectivity achieved for the entire mirror is mostly determined by the ML, including ML pair numbers and the refractive index contrast between the composed materials. It can be seen in figure 2.7(a) that the refractive indexes between Mo and Si over the wavelengths have differences, which affects only the refractive angle of EUV. In contrast, the extinction coefficients between Mo and Si have huge differences, which means that

Table 2.1. Thicknesses of Ml, Si, Ru, Cr and the Si substrate used to calculate figure 2.6.

Material	Si	Mo	Substrate (Si)	Capping (Ru)	Backside coating (Cr)
Thickness in *z*-direction	4.17 nm	2.78 nm	Infinite thick	2 nm	2 nm

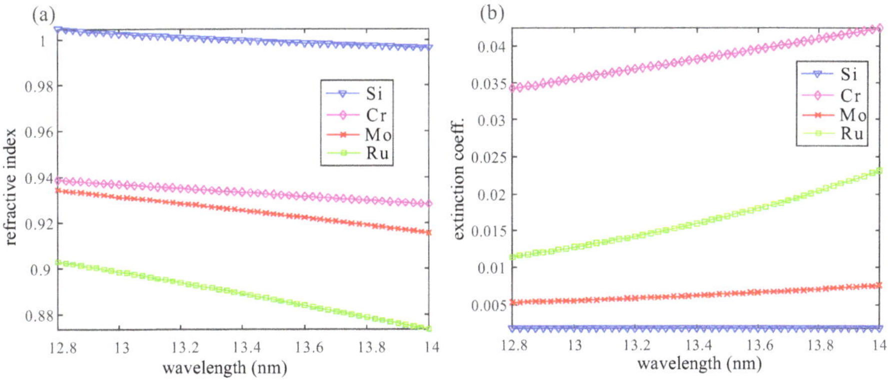

Figure 2.7. Refractive indexes and extinction coefficients of Mo, Si, Cr and Ru.

the EUV is absorbed more in Mo than in Si. Therefore, the Mo serves as an absorber in the Mo/Si ML. As shown in figure 2.6(a), the reflectivity that the Mo/Si based ML mirror undergoes in the EUV wavelength ranging from 12.8 nm to 14 nm illumination was calculated. Apparently, for both polarizations, the peak reflectivity appears at CW = 13.5 nm as it is now the standard central wavelength used in the EUV lithography system. By illuminating 13.5 nm EUV onto the Mo/Si based ML mirror, the reflectivity variation as a function of Mo/Si pair number is calculated in figure 2.6(b). From the plot readers can see that, before reflectivity reaches 70%, it increases rapidly for both curves by adding layers. Owing to the nature of material extinction, the reflectivity increment becomes very slow after 70% and eventually it will be saturated. The calculated reflectivity curves versus incident angles of 40 pairs of Mo/Si based ML mirror undergoing 13.5 nm EUV illumination can be seen in figure 2.6(c). Both curves reveal that the peak reflectivity occurs at an incident angle of 6°, this default angle of EUV photomask is the main stream in the EUV lithography system [39]. From the results we see that EUV ML mirrors have a very narrow transparent band, the variation is on the order of 1 nm, requiring EUV optics to be specifically matched to the wavelength of the source.

2.4 Quarter-wavelength EUV mirror (Bragg mirror)

A common mirror design used in EUV applications is the ML based quarter-wave mirror, also called a Bragg mirror. The concept of the quarter-wave mirror is based on a dielectric structure that consists of an alternating sequence of layers of two different optical materials corresponding to alternate high- and low-index. These two different materials cause a specific band of wavelengths to constructively interfere and reflect. Part of the incident beam is reflected at each interface in the stack. For which quarter-wave mirror is designed, each optical layer thickness in the mirror corresponds to one-quarter of the applied wavelength within each layer. For instance, a periodic Mo/Si based ML as shown in figure 2.5(a). Assuming Mo and Si are both absorption-free to the applied wavelength, the alternative Mo/Si layers can be fabricated with appropriate thickness corresponding to the applied wavelength λ and illuminated angle to let the optical path length be equal to quarter-wavelength, which is

$$n_{\text{Mo}}T_{\text{Mo}}\cos\theta_{\text{Mo}} = n_{\text{Si}}T_{\text{Si}}\cos\theta_{\text{Si}} = \frac{\lambda}{4}, \tag{2.37}$$

where n_{Mo} and n_{Si} are real part refractive indexes of Mo and Si. If this is the case, then the phase angle of both quarter-wavelength thickness Mo and Si films becomes

$$\beta_{\text{Mo}-Q} = \beta_{\text{Si}-Q} = \frac{\pi}{2}. \tag{2.38}$$

So the corresponding characteristic matrix, which is from equation (2.32), can be simplified as

$$M_{\text{Mo/Si}}^{Q} = \begin{bmatrix} -\dfrac{g_{\text{Si}}}{g_{\text{Mo}}} & 0 \\ 0 & -\dfrac{g_{\text{Mo}}}{g_{\text{Si}}} \end{bmatrix}. \tag{2.39}$$

For a N period Mo/Si with quarter-wavelength thickness, the characteristic matrix becomes

$$M_{\text{Mo/Si}}^{N,Q} = \begin{bmatrix} \left(-\dfrac{g_{\text{Si}}}{g_{\text{Mo}}}\right)^{N} & 0 \\ 0 & \left(-\dfrac{g_{\text{Mo}}}{g_{\text{Si}}}\right)^{N} \end{bmatrix}. \tag{2.40}$$

Based on the one-dimensional analytical solution, the reflection and transmission coefficients of a Mo/Si mirror which consists of N period quarter-wave thickness Mo/Si pair are

$$r_{\text{Mo/Si}}^{N,Q} = \frac{1 - \dfrac{g_{\text{sub}}}{g_{\text{vacuum}}}\left(\dfrac{g_{\text{Mo}}}{g_{\text{Si}}}\right)^{2N}}{1 + \dfrac{g_{\text{sub}}}{g_{\text{vacuum}}}\left(\dfrac{g_{\text{Mo}}}{g_{\text{Si}}}\right)^{2N}} \tag{2.41}$$

and

$$t_{\text{Mo/Si}}^{N,Q} = \frac{-2\left(\dfrac{g_{\text{Mo}}}{g_{\text{Si}}}\right)^{N}}{1 + \dfrac{g_{\text{sub}}}{g_{\text{vacuum}}}\left(\dfrac{g_{\text{Mo}}}{g_{\text{Si}}}\right)^{2N}}. \tag{2.42}$$

Broadly speaking, the principle of operation within the quarter-wavelength Mo/Si mirror can be understood as follows. Each interface between the Mo/Si contributes a

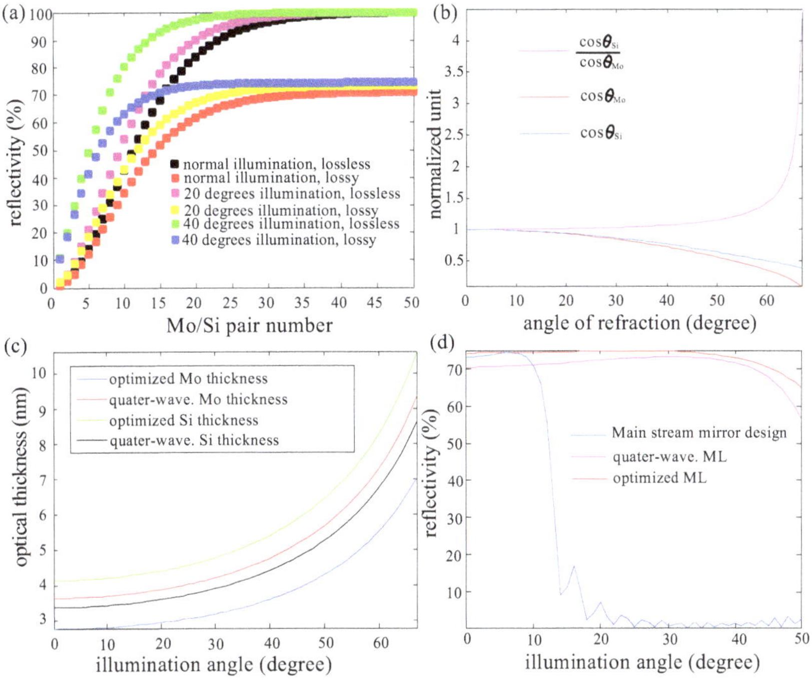

Figure 2.8. (a) Reflectivity as a function of pair numbers of Mo/Si ML based quarter-wavelength mirrors, in which reflectivities of a 13.5 nm EUV light illuminates from different angles onto the lossless and the lossy mirrors are calculated. (b) Cosine values of quarter-wavelength designed Mo and Si layers as a function of angle of refraction under 13.5 nm EUV illumination. (c) Comparisons between the optimized Mo/Si layer thicknesses and the quarter-wavelength designed thicknesses for different EUV illuminations. (d) Reflectivity of 40 Mo/Si ML based mirrors as a function of 13.5 nm EUV illumination angles, in which the main stream design, the quarter-wavelength design and the optimized designs are plotted. In the plots only *s*-polarization was calculated.

Fresnel reflection. Nevertheless, the optical path length difference between reflections from subsequent interfaces is half the wavelength. The Fresnel reflections have a phase shift of 180° only if the incident light goes from low-index medium in a high-index medium. Consequently, the relative phase difference of all reflected beams is 0° or a multiple of 360°. Therefore, all reflected components from the interfaces interfere constructively, which results in a strong reflection.

Based on equation (2.42), the reflectivities of Mo/Si ML based quarter-wavelength mirror were calculated as a function of Mo/Si pair number in figure 2.8(a). We considered both lossy (i.e. apply extinction coefficient in complex refractive index) and lossless (i.e. remove extinction coefficient from complex refractive index) conditions with different EUV illuminations. The refractive index that is used for the lossless Mo, Si and Cr are n_{Mo}, n_{Si}, and n_{Cr} at 13.5 nm, the readers can refer to figure 2.7(a) for the values. For the lossy mirror, the complex refractive index at 13.5 nm is used, which can be expressed as $\tilde{n}_{\mathrm{Mo}} = n_{\mathrm{Mo}} - \mathrm{j}k_{\mathrm{Mo}}$, $\tilde{n}_{\mathrm{Si}} = n_{\mathrm{Si}} - \mathrm{j}k_{\mathrm{Si}}$, and $\tilde{n}_{Cr} = n_{\mathrm{Cr}} - \mathrm{j}k_{\mathrm{Cr}}$. The readers can also refer to figures 2.7(a) and (b) for the values.

The thickness of the substrate, the capping, and the backside coating layers are shown in table 2.1.

As shown in figure 2.8(a), for the lossless Mo/Si ML based quarter-wavelength mirror, the reflectivities under EUV illuminations are able to approach 100% with the increment of pair number N. The result matches the expression in equation (2.41) since $n_{\mathrm{Mo}} < n_{\mathrm{Si}}$. In addition, the response time for reflectivity to approach 100% is faster for larger illumination angle due to the faster increment of the ratio of $\cos\theta_{\mathrm{Si}}$ to $\cos\theta_{\mathrm{Mo}}$. Figure 2.8(b) shows the trend of cosine values of refractive angles for 13.5 nm EUV light travels through Mo and Si. It is obvious, from figure 2.8(a), that the reflectivity under larger illuminated angle will approach the saturated reflectivity with fewer Mo/Si pairs. The thing is that it is not possible for a lossy Mo/Si ML based quarter-wavelength mirror to reach 100% reflectivity due to material extinction. In figure 2.8(a), for all illumination angles, the reflectivities of the lossy Mo/Si ML based quarter-wavelength mirror approach saturated reflectivity after using more than 25 Mo/Si pairs. Mostly, the saturated reflectivity for all illuminations approach only roughly 70%, even adding more Mo/Si pairs.

Due to the absorptions of Mo and Si, quarter-wavelength stack designs based on equation (2.37) are no longer the best way to achieve the highest reflectivity. For the analysis of light travel through the lossy material, the complex refractive angle has been derived in [40], which means when absorption cannot be neglected, the optimum value of layer thickness may be different from quarter-wave stack design [41]. A transcendental equation for the absorption layer under normal incidence has been derived in [42], and the choice of layer thickness for higher reflectivity is

$$\tan\pi\left(\frac{T_{\mathrm{Mo}}}{T_{\mathrm{Mo}}+T_{\mathrm{Si}}}\right)=\pi\left[\left(\frac{T_{\mathrm{Mo}}}{T_{\mathrm{Mo}}+T_{\mathrm{Si}}}\right)+\left(\frac{\varphi_{\mathrm{Si}}}{\varphi_{\mathrm{Mo}}-\varphi_{\mathrm{Si}}}\right)\right] \tag{2.43}$$

with

$$\varphi_{\mathrm{Mo}}=\frac{k_{\mathrm{Mo}}\lambda}{4\pi} \text{ and } \varphi_{\mathrm{Si}}=\frac{k_{\mathrm{Si}}\lambda}{4\pi}. \tag{2.44}$$

The unconstrained mathematical model for the reflectivity based on 40 periods Mo/Si is used to find the thickness of the Mo and Si for the higher reflectivity. To proceed this, below one-dimensional minimization is used to find the minimum of $-r_{\mathrm{Mo/Si}}^{N=40}$ within a thickness interval of Mo and Si:

Minimize:

$$\text{minimize}_{T_{\mathrm{Mo}},\,T_{\mathrm{Si}}}-r_{\mathrm{Mo/Si}}^{N=40}(T_{\mathrm{Mo}},\,T_{\mathrm{Si}}), \tag{2.45}$$

$$\text{Subject to: }\begin{cases}T_{\mathrm{Mo}}^{\min}\leqslant T_{\mathrm{Mo}}\leqslant T_{\mathrm{Mo}}^{\max}\\ T_{\mathrm{Si}}^{\min}\leqslant T_{\mathrm{Si}}\leqslant T_{\mathrm{Si}}^{\max}\end{cases}$$

where $T_{\mathrm{Mo}}^{\min}$ and $T_{\mathrm{Mo}}^{\max}$ stand for the constrain boundaries of T_{Mo}. $T_{\mathrm{Si}}^{\min}$ and $T_{\mathrm{Si}}^{\max}$ stand for the constrain boundaries of T_{Si}. By applying equation (2.45) in one-dimensional analytical solution, the 40 pairs Mo and Si thicknesses of higher mirror reflectivity for different EUV illuminations were optimized. The thickness variation under different EUV illuminations are plotted together with the quarter-wavelength design in figure 2.8(c). What we can find, after optimization, in figure 2.8(c) are that the optimized Mo thickness become thinner and the optimized Si thickness become thicker than the quarter-wavelength design over all EUV illumination angles. The physical meaning is that the Mo layers contribute more EUV absorptions than Si layers by nature, thus the minimization function minimized all the Mo thickness to reduce the entire EUV absorption of the mirror. Also, to preserve the required phase-shift, the thickness of Si layers need to be increased. The calculated mirror reflectivity of the optimized mirrors at all the illumination angles, quarter-wavelength mirror and the main stream mirror design (Mo: 2.78 nm, Si: 4.17 nm, used to calculated figure 2.6) are plotted in figure 2.8(d). For the main stream mirror design, the reflectivity varies from 70% to 74% within 10° EUV illumination, but its reflectivity drops down drastically after this region. On the other hand, the reflectivity of quarter-wavelength stack is about 70% within 10° EUV illumination, which implies the performance within this region is poorer than the main stream mirror design. Even so, it has wider bandwidth than the main stream mirror design. The reflectivity of the quarter-wavelength mirror maintains above 70% till 45° EUV illumination. For the optimized mirrors at all the illumination angles, they produce not only higher reflectivity but also wider bandwidth as compared to the other two, which use fixed thickness at all illumination angles. The reflectivity maintains around 74% from 0° to 45° EUV illumination. From figure 2.8(d), the optimized performance can be found at different EUV illuminations.

2.5 Photonic crystal (PHC) EUV mirror

PHC mirrors can be formed by composing matters in different form of periodicity depending on the desired band. Many artless photonic crystals can also be found throughout the natural world, the patterns on a butterfly's wings is a perfect example. Although PHCs seem to have existed for millions of years, humans have only recently started to explore their potential.

2.5.1 Photonic crystals (PHC)

The concept of PHC was proposed by E Yablonovitch [43] and S John [44]. PHCs stand for a structure made of periodic dielectrics, so that they contain regularly repeating regions of high and low refractive index. Photonic crystals often comprise multiple layers of different materials arranged, multilayered material can also be periodic in a one- two-, or three-dimensional (3D) arrangements as illustrated in figure 2.9. One-dimensional (1D) PHCs consist of stacks of identical parallel planar ML segments, the ML structure of EUV mirror is an example. 2D PHCs are composed of sets of parallel rods together with sets of parallel cylindrical holes. Three-dimensional PHCs include arrays of cubes, spheres, or holes of various shapes, organized in lattice structures.

Optical waves interact with PHCs in a unique way because the light waves are periodic by its nature, particularly when the scale of the periodicity is of the same

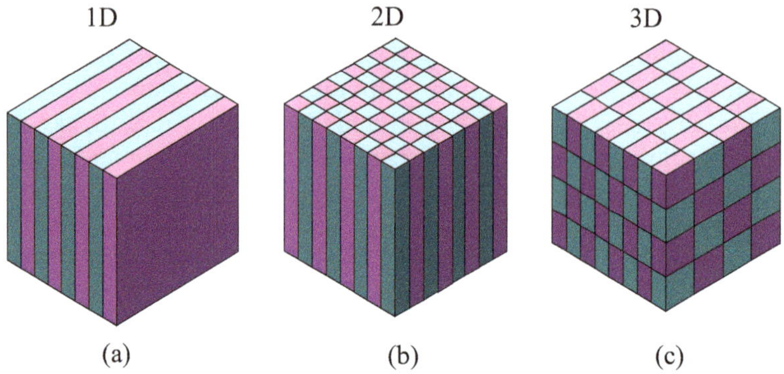

Figure 2.9. Geometrical shapes of photonic crystal structures in different crystal arrangements, in which (a) is periodic in one direction, (b) is periodic in two directions, and (c) is periodic in three directions.

order as that of the wavelength. For instance, well-designed EUV mirrors are able to reflect ~70% EUV at 13.5 nm. 13.5 nm EUV lying within these forbidden bands, called photonic stop bands or photonic bandgaps, behaves in a manner akin to ~70% reflection. Remarkably, this effect extends over finite spectral bands instead of just single frequency. Since PHCs affect the motion of photons in much the same way that ionic lattices affect electrons in solids, thus the photonic bands can be described just like the concept of electronic band structure.

2.5.2 Bloch theorem

The energy band theory, in solid state physics for an electron, was developed based on the research of applying quantum mechanics to periodic structure. The Schrödinger equation expresses the wave function of the energy band theory of the electron, the solution in one-dimensional space can be described as below according to Bloch's theorem [45]

$$\Psi(\vec{r}) = u(\vec{r})e^{-j\vec{K}\vec{r}}, \tag{2.46}$$

where $u(\vec{r})$ and $\vec{K}$ stand for the Bloch function and Bloch wave vector. For a periodic structure with period of Π, the Bloch function becomes

$$u(\vec{r}) = u(\vec{r} + \vec{\Pi}). \tag{2.47}$$

Owing to the similar behaviors of electromagnetic (EM) waves propagate in periodic structures and electrons act in crystalline solids, we are able to draw an analogy between equation (2.47) and the EM waves. Assuming we have a one-dimensional PHC with period of Π along the x-direction, both permittivity and permeability have the below periodicity

$$\varepsilon = \varepsilon(x + \Pi) \tag{2.48}$$

and

$$\mu = \mu(x + \Pi). \tag{2.49}$$

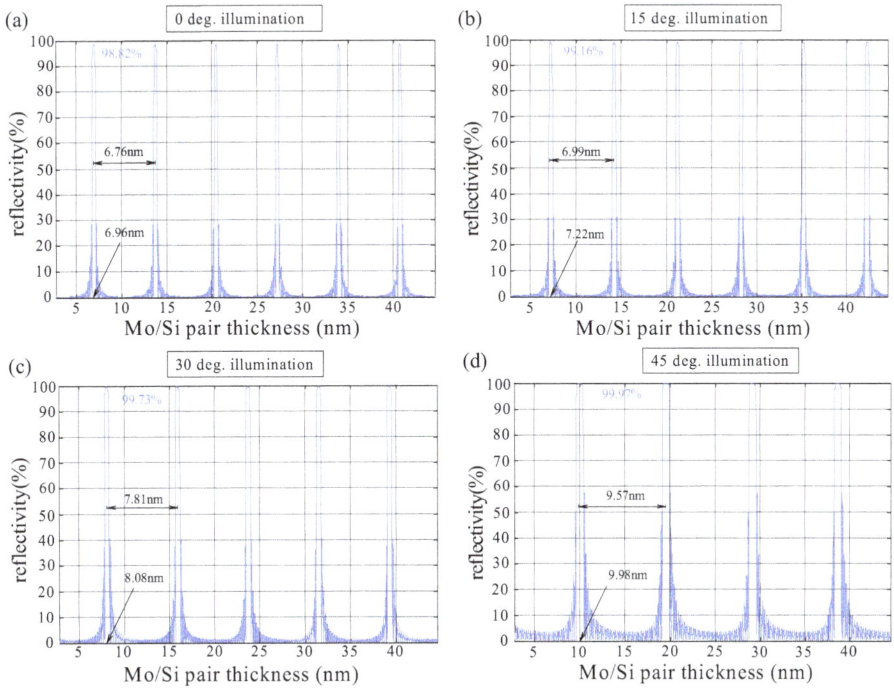

Figure 2.10. Calculated EUV stop and pass bands of a 40 pairs loss-free Mo/Si ML based mirror, in which the EUV is illuminated at an angle of (a) 0° (b) 15° (c) 30° (d) 45°.

According to the Bloch theorem, therefore, both electric and magnetic fields also have the below periodicity

$$E(x + \Pi) = |\vec{E}(x)e^{-j\vec{K}\vec{\Pi}}| \tag{2.50}$$

and

$$H(x + \Pi) = |\vec{H}(x)e^{-j\vec{K}\vec{\Pi}}|. \tag{2.51}$$

This periodicity should also be applied to stop band and pass band. By using the one-dimensional analytical solution derived in section 2.3.2, the EUV bands as a function of the spatial period of 40 pairs loss-free Mo/Si ML based mirror are calculated in figure 2.10, both stop bands and pass bands are found under different EUV illuminations. The loss-free refractive index of Mo and Si are 0.9238 and 0.999, the thickness of Mo and Si are 2.78 nm and 4.17 nm. In figure 2.10, only *s*-polarization is plotted since *p*-polarization will be similar. From the results in figures 2.10(a)–(d), both stop and pass bands under different illuminations have their periodicity. As seen from figure 2.10(a), when the EUV illuminates loss-free Mo/Si ML based mirror with 0° illumination, the period Π of both EUV stop and pass bands is 6.76 nm. When we enlarge this illumination angle shown in figure 2.10(b)–(d), the period Π and the bandwidth of both stop and pass bands are increased as well. From the view point of mirror reflectivity, under 0° illumination, the peak value of EUV stop band limits at 98.82% as shown in figure 2.10(a). When we increase the illumination angle to 15°, 30° and 45° as shown in figures 2.10(b)–(d), the EUV stop band keeps improving to 99.16%, 99.73% and 99.97% but not too much since 98.82% is already strong. Besides, the increment of illumination angle also weaken the pass band. From a mirror design perspective, the small illumination angle has the capacity to design the thinner mirror thickness as compared with large illumination angle. This can be observed from the pair thickness of the first stop band in figures 2.10(a)–(d). As shown in figure 2.10(a), when applying 0° illumination, the first Mo/Si pair thickness we can use to design a mirror with 98.82% reflectivity is 6.96 nm. However, if the illumination angle is increased to 15°, 30° and 45° as shown in figures 2.10(b)–(d), the Mo/Si pair thickness of the first stop band becomes 7.22 nm, 8.08 nm and 9.98 nm. For a mirror with 40 layer design, the entire ML thickness will be increased from 278 nm to 399 nm, which is quite a thickening (~44%).

2.5.3 Quasi-two-dimensional PHC EUV mirror

From section 2.4 and our literature reviews, which are described in section 2.3, the reflectivity of the EUV mirrors have not had many improvements owing to the unavoidable loss tangent of the composed materials. This triggers the idea to decrease the material loss of the mirror by arranging low-loss substance or hollow crystals into it. As shown in figure 2.11(a), we arrange hollow cylinders with infinite length along the *y*-direction in each Si layer of the 40 pairs Mo/Si based EUV mirror. The purpose is to reduce the loss tangent of each Si layer. By doing this, we turn a Mo/Si based mirror into a 2D photonic crystal structure. However, to calculate the bands of this structure by using the 1D analytical solution which fits only for 1D structure, the equivalent index of the Si-hollow mixed layer Si_{eq} needs to be used, as shown in figure 2.11(b). The equivalent indices can be calculated from the effective-medium theory [46]. In addition, we also reduce the loss tangent of each Mo layer by doing the same thing. The equivalent index of the Mo-hollow mixed layer is Mo_{eq}, which can be seen in figure 2.11(c). The equivalent indices of both Si_{eq} and Mo_{eq} films under different Si-hollow and Mo-hollow mix ratios are listed in table 2.2.

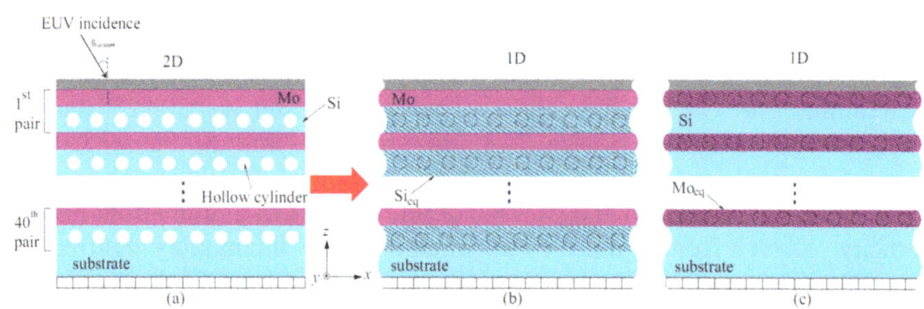

Figure 2.11. Illustrations of an EUV light illuminates onto (a) a 40 pairs Mo/Si ML based 2D mirror with several hollow cylinders embeds in Si layers, (b) a 40 pairs Mo/Si_{eq} ML based 1D mirror and (c) a 40 pairs Mo_{eq}/Si ML based 1D mirror.

Table 2.2. Equivalent refractive indices and equivalent extinction coefficients of Mo and Si layers under different hollow-Mo or hollow-Si mix ratios.

Hollow-Si/Mo mix ratio (%)	Equivalent refractive indices n_{eq}		Equivalent extinction coefficients k_{eq}	
	Si_{eq}	Mo_{eq}	Si_{eq}	Mo_{eq}
0	0.9990	0.9238	0.0018	0.0064
10	0.9991	0.9312	0.0016	0.0058
20	0.9992	0.9386	0.0015	0.0052
30	0.9993	0.9460	0.0013	0.0046
40	0.9994	0.9536	0.0011	0.0040
50	0.9995	0.9612	0.0009	0.0033
60	0.9996	0.9688	0.0007	0.0027
70	0.9997	0.9765	0.0005	0.0020
80	0.9998	0.9843	0.0004	0.0014
90	0.9999	0.9921	0.0002	0.0007

By applying these equivalent indices in one-dimensional analytical solution derived in section 2.3.2, EUV (λ = 13.5 nm) bands as a function of the spatial period of 40 pairs Mo/Si, Mo/Si_{eq} and Mo_{eq}/Si ML based mirrors are calculated in figure 2.12 (*s*-polarization). Here the illumination angle of EUV is 6°, the dimension of each layer can be seen in table 2.1. As shown in figure 2.12(a), owing to the effect of the extinction coefficient of the solid Mo/Si based mirror, the stop bands no longer reaches 99%. Instead, the highest stop band limits at 74.25% with the thinnest Mo/Si pair thickness, and it keeps going down with the increment of the pair thickness. This phenomenon makes sense since increasing pair thickness intensifies the effect of the extinction coefficient (i.e. both spacer and absorber are getting thicker, light transmission will become harder). Anyhow, for solid Mo/Si based mirror, both stop and pass bands still preserve the periodicity as described in Bloch's theorem. For figures 2.12(b)–(d), they are the bands comparison between the Mo/Si_{eq} based mirror and Mo_{eq}/Si based mirror for the ratios of 30%, 60% and 90%. As seen from figure 2.12(b), as expected, all the stop bands go down with the increment of the pair thickness because of the residual 70% extinction coefficient. The stop bands of Mo/Si_{eq} based mirror are intensified as compared to solid Mo/Si based mirror. In addition, the pair thickness of the first stop band is 2 nm thinner than the solid Mo/Si (from 6.95 nm to 6.93 nm) which is also a plus. In contrast, the stop bands of Mo_{eq}/Si based mirror are weakened. This phenomenon can be explained through table 2.2, although the extinction coefficient of the Mo_{eq} does decrease from 0.0064 to 0.0046 when it becomes 30% hollow out, the refractive index of Mo_{eq} becomes close to Si_{eq}. Thus, the reflection from the Mo_{eq}/Si based mirror becomes weakened. Moreover, the phenomenon similar to figure 2.12(b) with more obvious contrast between Mo_{eq}/Si and Mo/Si_{eq} based mirrors can be seen in figures 2.12(c) and (d). By increasing the hollow-out percentage in Si, the stop bands can be intensified, and the effect of the

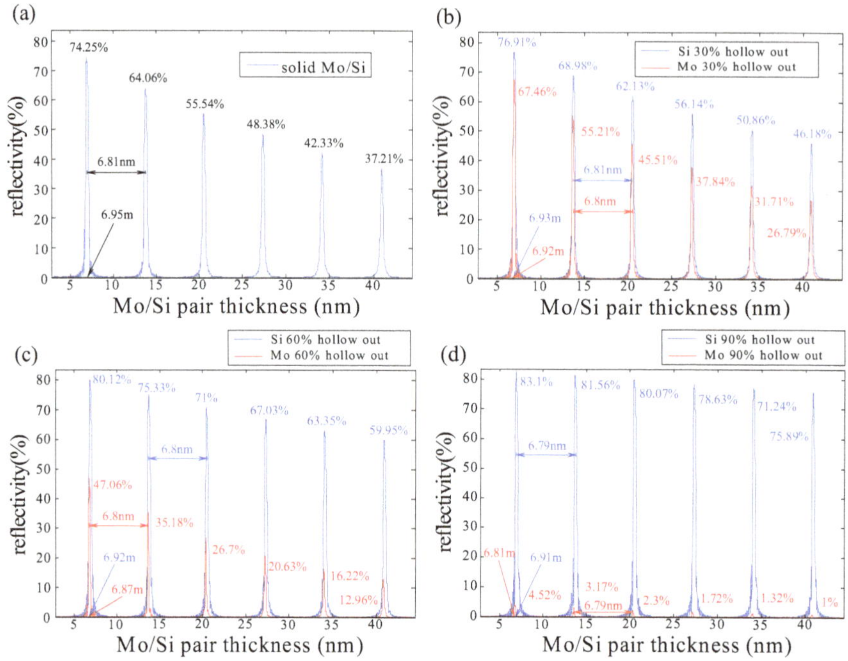

Figure 2.12. Calculated EUV stop and pass bands of (a) a 40 pairs Mo/Si ML based mirror, (b) a 40 pairs Mo/Si ML based mirror with either 30% Mo or Si hollow out, (c) a 40 pairs Mo/Si ML based mirror with either 60% Mo or Si hollow out and (d) a 40 pairs Mo/Si ML based mirror with either 90% Mo or Si hollow out under 6° EUV illumination.

extinction due to pair thickness increment can be reduced. Also, the pair thickness of the first stop band keeps thinning. In contrast, by increasing the hollow-out percentage in Mo, the stop bands are weakened, and it does not improve the extinction due to pair thickness increment. To sum up, to design a high reflectivity EUV mirror, there are pros to decrease material loss of each Si layer by arranging hollow crystals or low-loss substance into it, and there are cons if the designer applies the same thing to each Mo layer.

References

[1] Spiller E 1994 Imaging systems for x-rays *Soft X-ray Optics* (Bellingham, WA: SPIE) ch 4

[2] Levinson H J 2010 Wafer steppers *Principles of Lithography* 3rd edn (Bellingham, WA: SPIE) ch 5

[3] Okoroanyanwu U 2010 X-ray and extreme ultraviolet lithographies *Chemistry and Lithography* (Bellingham, WA: Wiley) ch 14

[4] SEMI (https://semi.org/zh/products-services/standards/using-semi-standards) (Accessed 19 September 2020)

[5] Yan P Y 2005 Masks for extreme ultraviolet lithography *Handbook of Photomask Manufacturing Technology* 1st edn ed S Rizvi (Boca Raton, FL: CRC Press) ch 11

[6] Jonckheere R, Lorusso G F, Goethals A, Ronse K, Hermans J and Ruyter R D 2007 Assessment of EUV reticle blank availability enabling the use of EUV tools today and in the future *Proc. SPIE* **6533** 653313
[7] Peters J H 2005 Status of EUVL mask development in Europe *Proc. SPIE* **5853** 297–307
[8] Hector S D and Kemp K 2004 EUVL mask challenges and how International SEMATECH is addressing them *Proc. SPIE* **5446** 792–803
[9] Hayashi K 2007 Development status of EUVL mask blanks in AGC *Proc. SPIE* **6730** 67305D
[10] Voorma H J, Louis E, Koster N B and Bijkerk F 1997 Fabrication and analysis of extreme ultraviolet reflection masks with patterned W/C absorber bilayers *J. Vac. Sci. Technol.* B **15** 293–8
[11] Yan P Y, Zhang G, Kofron P, Chow J, Stivers A, Tejnil E, Cardinale G and Keamey P 1999 EUV mask patterning approaches *Proc. SPIE* **3676** 309–13
[12] Yan P Y, Zhang G, Kofron P, Powers J E, Tran M, Liang T, Stivers A R and Lo F C 2000 EUV mask absorber characterization and selection *Proc. SPIE* **4066** 124–30
[13] Mangat P J S, Hector S D, Rose S, Cardinale G F, Tejnil E and Stivers A R 2000 EUV mask fabrication with Cr absorber *Proc. SPIE* **3997** 76–82
[14] Wasson J R, Smith K H, Mangat P J S and Hector S D 2001 Infinitely selective repair buffer for EUVL reticles *Proc. SPIE* **4343** 402–8
[15] Lee B T *et al* 2001 Characteristics of the Ru buffer layer for EUVL mask patterning *Proc. SPIE* **4343** 746–53
[16] Zhang G *et al* 2007 EUV mask process development and integration *Proc. SPIE* **6283** 62830G
[17] Shoki T, Hosoya M, Kinoshita T, Kobayashi H, Usui Y, Ohkubo R, Ishibashi S and Nagarekawa O 2002 Process development of 6-in EUV mask with TaBN absorber *Proc. SPIE* **4754** 857–64
[18] Takahashi M, Ogawa T, Hoko H, Yamanashi H, Hirano N, Chiba A, Ito M and Okazaki S 2000 Smooth low-stress sputtered tantalum and tantalum alloy films for the absorber material of reflective-type EUVL *Proc. SPIE* **3997** 484–95
[19] Yan P Y, Zhang G, Ma A and Liang T 2001 TaN EUVL mask fabrication and characterization *Proc. SPIE* **4343** 409–14
[20] Han S I, Wasson J R, Mangat P J S, Cobb J L, Lucas K and Hector S D 2002 Novel design of att-PSM structure for extreme-ultraviolet lithography and enhancement of image contrast during inspection *Proc. SPIE* **4688** 481–94
[21] Hoshino E, Ogawa T, Takahashi M, Hoko H, Yamanashi H, Hirano N and Okazaki S 1999 Damage control during dry etching of EUV mask: I. Control of surface roughness *Proc. SPIE* **3873** 786–91
[22] Yan P Y 2002 EUVL alternating phase shift mask imaging evaluation *Proc. SPIE* **4889** 1099–105
[23] Yulin S, Benoit N, Feigl T, Kaiser N, Fang M and Chandhok M 2008 Mo/Si multilayers with enhanced TiO2- and RuO2-capping layers *Proc. SPIE* **6921** 692118
[24] Zhang G *et al* 2006 EUV mask process development and integration *Proc. SPIE* **6283** 62830G
[25] Attwood D 1999 Introduction *Soft X-rays and Extreme Ultraviolet Radiation: Principle and Applications* 1st edn (Cambridge: Cambridge University Press) ch 1

[26] Lee Y M, Li J H, Ng P C W, Pei T H, Wang F M, Tsai K Y and Chen A C 2010 Efficient scattering simulations for equivalent extreme ultraviolet mask multilayer structures by modified transmission line theory and finite-difference time-domain method *J. Micro/ Nanolith. MEMS MOEMS* **9** 043003

[27] Yamaguchi T, Ikuta H, Tomofuji T, Yanagi Y, Itoh Y, Oka T and Mizutani U 2008 Reflective properties of Mo–Si multilayer for EUV lithography deposited by the magnetron sputtering device with superconducting bulk magnets *Physica* C **468** 2170–3

[28] Hiruma K, Miyagaki S, Yamanashi H, Tanaka Y and Nishiyama I 2008 Performance and quality analysis of Mo-Si multilayers formed by ion beam and magnetron sputtering for extreme ultraviolet lithography *Thin Solid Films* **516** 2050–7

[29] Yoshioka M *et al* 2009 Xenon DPP source technologies for EUVL exposure tools *Proc. SPIE* **7271** 727109

[30] Ronse K, Jansen P, Gronheid R, Hendrickx E, Maenhoudt M, Wiaux V, Goethals A M, Jonckheere R and Vandenberghe G 2009 Lithography options for the 32 nm half pitch node and beyond *IEEE Trans. Circuit Sys.* **56** 1884–91

[31] Carniglia C K and Apfel J H 1980 Maximum reflectance of multilayer dielectric mirrors in the presence of slight absorption *J. Opt. Soc. Am.* **70** 523–34

[32] Singh M and Braat J J M 2000 Design of multilayer extreme-ultraviolet mirrors for enhanced reflectivity *Appl. Opt.* **39** 2189–97

[33] Kang I Y, Kim T G, Lee S Y, Ahn J and Chung Y C 2004 Structural characterization of a Mo/Ru/Si extreme ultraviolet (EUV) reflector by optical modeling *Jpn. J. Appl. Phys.* **43** 3700–2

[34] Modi M H, Rai S K, Idir M, Schaefers F and Lodha G S 2012 NbC/Si multilayer mirror for next generation EUV light sources *Opt. Express* **20** 15114–20

[35] Maury H, André J M, Guen K L, Mahne N, Giglia A, Nannarone S, Bridou F, Delmotte F and Jonnard P 2009 Analysis of periodic Mo/Si multilayers: Influence of the Mo thickness *Surf. Sci.* **603** 407–11

[36] Suman M, Pelizzo M G, Nicolosi P and Windt D L 2008 Aperiodic multilayers with enhanced reflectivity for extreme ultraviolet lithography *Appl. Opt.* **47** 2906–14

[37] Pelizzo M G, Suman M, Monaco G, Nicolosi P and Windt D L 2008 High performance EUV multilayer structures insensitive to capping layer optical parameters *Opt. Express* **16** 15228–37

[38] The Center for X-Ray Optics (CXRO) (http://cxro.lbl.gov/) (Accessed 19 September 2020)

[39] Capelli R, Garetto A, Magnusson K and Scherüb T 2014 Scanner arc illumination and impact on EUV photomasks and scanner imaging *Proc. SPIE* **9231** 923109

[40] Harrington R F 2001 Introduction to waves *Time-Harmonic Electromagnetic Fields* 1st edn (New York: IEEE Press) ch 2

[41] Underwood J H and Barbee T W 1981 Layered synthetic microstructures as Bragg diffractors for x-rays and extreme ultraviolet: theory and predicted performance *Appl. Opt.* **20** 3027–34

[42] Vinogradov A V and Zeldovich B Y 1977 X-ray and far uv multilayer mirrors: principles and possibilities *Appl. Opt.* **16** 89–93

[43] Yablonovitch E 1987 Inhibited spontaneous emission in solid-state physics and electronics *Phys. Rev. Lett.* **58** 2059–62

[44] John S 1987 Strong localization of photonics in certain disordered dielectric supperlattices *Phys. Rev. Lett.* **58** 2486–9

[45] Charles K 1996 Crystal structure *Introduction to Solid State Physics* 8th edn ed S Johnson (Hoboken, NJ: Wiley) ch 1
[46] Wang Z Q, Stroud D and Dregia S A 1994 Effective-medium theory for the optical properties of diamond and diamond like films *Phys. Rev.* **B 50** 12073–7

Chapter 3

Efficient finite-difference time-domain (FDTD) approach

Extreme ultraviolet (EUV) mirror design requires rigorous simulations, the finite-difference time-domain (FDTD) approach is the mainstream numerical method for solving electromagnetic (EM) problems. The rigorous FDTD method is calculated based on Maxwell's equations, in general a more accurate result can be obtained by using finer meshes (or grids) throughout the entire computation. However, using fine meshes results in potential overflow issue owing to huge memory requirement and long computational time. Moreover, it is tough to calculate accurate results for an EUV mirror because of the multiple thin layers of the multilayer (ML) structure, which induce the computational complexity. By introducing an efficient FDTD approach into calculations, we are capable of treating the ML structure as an effective one-layer material, so that the calculation efforts can be drastically reduced while preserving the required accuracy. This chapter is organized as follows. In section 3.1 we will review numerical methods for EUV mirror design from published studies. In section 3.2, we will walk through the basics of the FDTD method. Following that, often used boundary conditions that are critical to the FDTD method will be introduced in section 3.3. At the end of this chapter, the efficient FDTD approach that is able to improve the efficiency from rigorous calculations will be presented.

3.1 Numerical methods for EUV mirror design

From the previous research we can see that different numerical methods have been applied not only to feasibility studies of mirrors but real design. In order to facilitate the study of the EUV mirror, the below numerical methods are commonly used, they are the rigorous coupled wave analysis (RCWA) method, the differential method (DIM), the waveguide method (WGM), the finite-element method (FEM), and the FDTD method [1–3]. RCWA, DIM and WGM utilize the Fourier function in the design approach. Thus, these three methods can be good tools for 2D design.

doi:10.1088/978-0-7503-2652-0ch3

However, when applying these three methods to solve 3D problems, the drawback is we have to suffer long computational time because of the slow convergence of Fourier expansions. On the other hand, the FDTD method and FEM are suitable to simulate the complex 3D structures. These two methods approach the problem by numerically solving Maxwell's equations in the differential and integral forms, which are two different categories: time domain and frequency domain approaches. From the literature reviews, these two methods have indeed been used to simulate the EUV ML mirror [4–6]. In this book, the FDTD method is chosen as the tool for mirror design owing to the below advantages:

- The calculation speed is relatively fast compared to the frequency domain methods, since no storage in time and no matrix inversion are needed.
- It returns both near-field and far-field solutions from a single execution.
- It provides high accuracy for different EM and photonics problems.
- It provides time-domain solutions.

To simulate the scattering effects of the EUV mirror by using FDTD method, the memory requirement and the computational time consumption mostly depend on the applied mesh numbers. The mesh numbers used in computational domain fluctuate according to the mesh size used for the problem, the optical wavelength (CW = 13.5 nm in our case), and the complexity of the scattering structure. Theoretically, more accurate results can be obtained by using finer meshes. However, the trade-offs are longer computational time and a massive memory consumption. In addition, for EUV mirror simulations, the meshes to be discretized by the FDTD method require more careful consideration to ensure both numerical accuracy and numerical stability of the scattering results, because of multiple thin layers within EUV mirrors.

3.2 FDTD expression for Maxwell's equation

3.2.1 Maxwell's equations

The FDTD method is a computational EM technique proposed by K S Yee in 1966 [7]. It is a powerful method of solving Maxwell's equations in all three dimensions and in time. The basics of the FDTD method are based on the discretizations of Maxwell's equations in both the time domain and the space domain, which are, Faraday's law for dispersive materials and can be written as

$$\frac{\partial \vec{B}}{\partial t} = -\nabla \times \vec{E}, \tag{3.1}$$

where $\vec{B}$ and $\vec{E}$ stand for magnetic flux density and electric field. Also, per Ampère's law

$$\frac{\partial \vec{D}}{\partial t} = \nabla \times \vec{H} - \vec{J} \tag{3.2}$$

where $\vec{H}$ is magnetic field intensity, $\vec{D}$ is electric flux density, $\vec{J}$ denotes current density, t is time. The details of Maxwell equations in the continuous domain transferred into the discrete domain can be seen in [8]. $\vec{B}$ and $\vec{D}$ are given by

$$\vec{B} = \mu_0 \vec{H} \tag{3.3}$$

and

$$\vec{D} = \varepsilon_0 \varepsilon_r \vec{E}, \tag{3.4}$$

where μ_0 is the permeability of free space, ε_0 is the permittivity of free space and ε_r is the relative permittivity of the medium. Equations (3.1) and (3.2) are formulated in the differential form which are mostly used in FDTD. Regardless, the integral form is sometimes used in FDTD. The integral form of (3.1) and (3.2) are

$$\oint \vec{E} dc = -\frac{\partial}{\partial t} \int \vec{B} \vec{s} dc, \tag{3.5}$$

and

$$\oint \vec{H} dc = -\frac{\partial}{\partial t} \int \vec{D} \vec{s} dc + \int \vec{J} \vec{s} dc. \tag{3.6}$$

Equations (3.5) and (3.6) are derived per Stokes' theorem [9], which integrates the exterior derivative dc over the whole circle c. $\vec{s}$ is the unit normal vector perpendicular to the surface of c.

3.2.2 Yee lattice

The illustration of a 3D Yee lattice in a Cartesian coordinate system can be seen in figure 3.1(a). The Yee lattice is a closed computational domain for FDTD calculation. Sources and scattering objects that are required for FDTD calculation are inside this computational domain. This domain is divided into several sub-spaces called cells. The differential form of Maxwell's equations which refers to equations (3.1) and (3.2) are approached in each cell, where Δx, Δy and Δz are the side lengths of the cells. By default, the Yee lattice is an enclosed space to EM waves. In other words, this domain is surrounded by perfect electric conductors (PEC), meaning it is like a sealed box that does not allow any leakage. Therefore, the only scenario we will see when EM scatterings touch the boundaries is back reflection. Still, the Yee lattice can also simulate the EM scatterings in open space. To calculate the numerical solution for an open space, the absorption boundary condition (ABC) needs to be applied, we will introduce this in section 3.3.1.

To calculate numerical solution by using the FDTD method based on the Yee lattice, the following steps can be used:

Time and space discretization: Equations (3.1) and (3.2) can be discretized in the form of either spatial domain or time domain or both, first-order difference is typically used to discretize the equation. However, higher order difference might be used if the design specification requires higher accuracy. However, the complexity and the stability will be trade-offs. The illustrations of different types of first-order differences along the x-direction can be seen in figure 3.1(b), they are forward difference, backward difference and central difference.

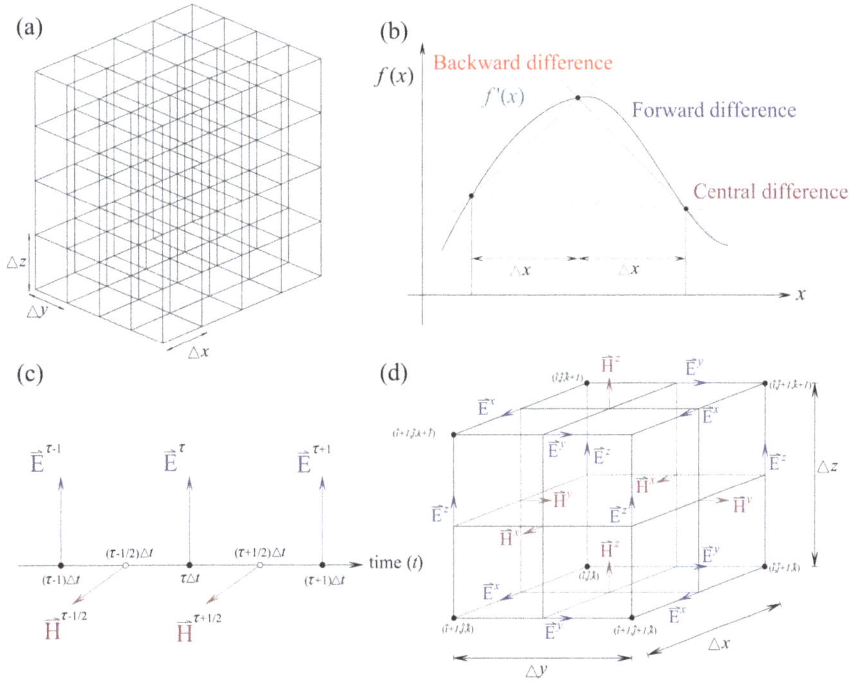

Figure 3.1. (a) 3D Yee lattice in Cartesian coordinate system. (b) Three finite-difference types, they are backward different, forward different and central difference. (c) Time arrangement of the EM fields in FDTD calculation. (d) 3D EM fields distribution on a single FDTD cell.

One thing to mention here is that the approximation error of both forward and backward differences is Δx, whereas the error of the central difference is $(\Delta x)^2$. Thus, the central difference is mostly applied in the FDTD method since it has higher accuracy compared to the other two. Take the electric field vector of the x-direction $\vec{E}^x$ as an example, the central difference formulation in time and spatial domains are

$$\frac{\partial \vec{E}^x}{\partial x} \approx \frac{\vec{E}^x\left(x + \frac{\Delta x}{2}, y, z, t\right) - \vec{E}^x\left(x - \frac{\Delta x}{2}, y, z, t\right)}{\Delta x} \tag{3.7}$$

and

$$\frac{\partial \vec{E}^x}{\partial t} \approx \frac{\vec{E}^x\left(x, y, z, t + \frac{\Delta t}{2}\right) - \vec{E}^x\left(x, y, z, t - \frac{\Delta t}{2}\right)}{\Delta t} \tag{3.8}$$

where x, y and z stand for the space axes in the 3D Cartesian coordinate system. Δt is the time step used for time domain calculations. Equations (3.7) and (3.8) can be converted into (3.9) and (3.10)

$$\frac{\partial \vec{E}^x}{\partial x} \approx \frac{\vec{E}^{x,\tau}\left(\hat{\imath} + \frac{1}{2}, \hat{\jmath}, \hat{k}\right) - \vec{E}^{x,\tau}\left(\hat{\imath} - \frac{1}{2}, \hat{\jmath}, \hat{k}\right)}{\Delta x}, \tag{3.9}$$

and

$$\frac{\partial \vec{E}^{x}}{\partial t} \approx \frac{\vec{E}^{x,\,\tau+\frac{1}{2}}\left(\hat{i},\hat{j},\hat{k}\right) - \vec{E}^{x,\,\tau-\frac{1}{2}}\left(\hat{i},\hat{j},\hat{k}\right)}{\Delta t}. \tag{3.10}$$

In equations (3.9) and (3.10), we convert the notation $\vec{E}^{x}(x, y, z, t)$ into $\vec{E}^{x,\,\tau}(\hat{i}, \hat{j}, \hat{k})$. Since the FDTD method divides both continuous space and time into discrete grids, (x, y, z, t) can be approximated by using $(\hat{i}\,\Delta x, \hat{j}\,\Delta y, \hat{k}\Delta z, \tau\Delta t)$.

Time arrangement of the EM fields: The EM waves are produced from constant interactions of the electric fields with magnetic fields, the continuous time system of the electric and magnetic fields are discretized by using central difference. After discretization, the time arrangement of the electric and the magnetic fields can be seen in figure 3.1(c). By applying equations (3.3) and (3.4), equations (3.1) and (3.2) becomes

$$\frac{\partial \vec{H}}{\partial t} = -\frac{1}{\mu_0}\nabla \times \vec{E}, \tag{3.11}$$

and

$$\frac{\partial \vec{E}}{\partial t} = -\frac{\sigma}{\varepsilon_0\varepsilon_r}\vec{E} + \frac{1}{\varepsilon_0\varepsilon_r}\nabla \times \vec{H} \tag{3.12}$$

where σ represents the conductivity. As seen from figure 3.1(c), the timing to allocate electric fields are $t = (\tau-1)\Delta t$, $(\tau)\Delta t$, $(\tau+1)\Delta t$, etc. On the other hand, the timing to allocate electric fields are $t = (\tau-1/2)\Delta t$, $(\tau+1/2)\Delta t$, etc. Therefore, the electric field at $t = (\tau-1/2)\Delta t$ and the magnetic field at $t = \tau\Delta t$ can be described through the central difference approximation

$$\frac{\partial \vec{E}}{\partial t} \approx \frac{\vec{E}^{\tau} - \vec{E}^{\tau-1}}{\Delta t}, \tag{3.13}$$

and

$$\frac{\partial \vec{H}}{\partial t} \approx \frac{\vec{H}^{\tau+\frac{1}{2}} - \vec{H}^{\tau-\frac{1}{2}}}{\Delta t}. \tag{3.14}$$

By substituting equations (3.13) and (3.14) for (3.11) and (3.12)

$$\frac{\vec{E}^{\tau} - \vec{E}^{\tau-1}}{\Delta t} = -\vec{E}^{\tau-\frac{1}{2}}\frac{\sigma}{\varepsilon_0\varepsilon_r} + \frac{1}{\varepsilon_0\varepsilon_r}\nabla \times \vec{H}^{\tau-\frac{1}{2}}, \tag{3.15}$$

and

$$\frac{\vec{H}^{\tau+\frac{1}{2}} - \vec{H}^{\tau-\frac{1}{2}}}{\Delta t} = -\frac{1}{\mu_0}\nabla \times \vec{E}^{\tau} \tag{3.16}$$

where $\vec{E}^{\tau-\frac{1}{2}}$ can be approximated by using $\frac{\vec{E}^{\tau} + \vec{E}^{\tau-1}}{2}$. So that we have

$$\vec{E}^{\tau} = \frac{1 - \frac{\sigma\Delta t}{2\varepsilon_0\varepsilon_r}}{1 + \frac{\sigma\Delta t}{2\varepsilon_0\varepsilon_r}}\vec{E}^{\tau-1} + \frac{\frac{\Delta t}{\varepsilon_0\varepsilon_r}}{1 + \frac{\sigma\Delta t}{2\varepsilon_0\varepsilon_r}}\nabla \times \vec{H}^{\tau-\frac{1}{2}}, \tag{3.17}$$

and

$$\vec{H}^{\tau+\frac{1}{2}} = \vec{H}^{\tau-\frac{1}{2}} - \frac{\Delta t}{\mu_0}\nabla \times \vec{E}^{\tau}. \tag{3.18}$$

From equations (3.17) and (3.18), we know that when doing FDTD calculations, $\vec{E}^{\tau}$ needs to be derived by using both $\vec{E}^{\tau-1}$ and $\vec{H}^{\tau-\frac{1}{2}}$. Accordingly, $\vec{H}^{\tau+\frac{1}{2}}$ requires both $\vec{E}^{\tau}$ and $\vec{H}^{\tau-\frac{1}{2}}$.

Spatial arrangement of the EM fields: The EM fields distribution can be derived from equation (3.9), with each FDTD cell illustrated in figure 3.1(d). The electric fields are distributed on the sidelines, where the magnetic fields are emitted from the center of each surface and are perpendicular to the surface. To sum up, the generic steps of a FDTD calculation are illustrated in figure 3.2. In the beginning we need to setup initial conditions such as EM wave source, mesh size, material properties, etc. Afterward, the

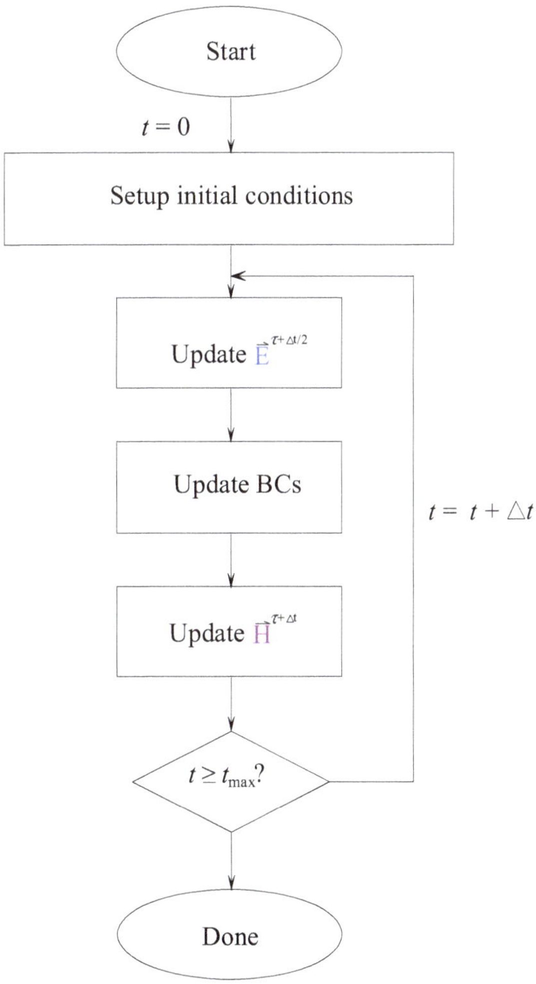

Figure 3.2. Illustration of the generic FDTD calculation flowchart.

first FDTD loop starts with the electric field calculation at $+\frac{\Delta t}{2}$. The next step is to update the boundary condition. Then, calculating the magnetic field for the next $\frac{\Delta t}{2}$. Thus, the first FDTD loop is completed, the next step is to check if the current timestamp reaches the time target. If the current timestamp is below the time target, the FDTD calculation needs to proceed for the next time step. In contrast, the FDTD calculation will be completed if the current timestamp reaches the time target.

3.2.3 FDTD expression in 3D

By referring to figure 3.1(d) together with the FDTD expressions of equations (3.17) and (3.18), the z-component of the electric field can be described as

$$\vec{E}^{z,\tau} = \frac{1-\dfrac{\sigma\Delta t}{2\varepsilon_0\varepsilon_r}}{1+\dfrac{\sigma\Delta t}{2\varepsilon_0\varepsilon_r}}\vec{E}^{z,\tau-1} + \frac{\dfrac{\Delta t}{\varepsilon_0\varepsilon_r}}{1+\dfrac{\sigma\Delta t}{2\varepsilon_0\varepsilon_r}}\left(\frac{\partial\vec{H}^{y,\tau-\frac{1}{2}}}{\partial x} - \frac{\partial\vec{H}^{x,\tau-\frac{1}{2}}}{\partial y}\right). \tag{3.19}$$

There are four $\vec{E}^z$ in each cell, if we use $\vec{E}^z\left(\hat{i}, \hat{j}, \hat{k}+\frac{1}{2}\right)$ in equation (3.19). The differential of $\vec{H}^x$ and $\vec{H}^x$ in equation (3.19) becomes $\frac{\vec{H}^{y,\tau-\frac{1}{2}}\left(\hat{i}+\frac{1}{2},\hat{j},\hat{k}+\frac{1}{2}\right)-\vec{H}^{y,\tau-\frac{1}{2}}\left(\hat{i}-\frac{1}{2},\hat{j},\hat{k}+\frac{1}{2}\right)}{\Delta x}$ and $\frac{\vec{H}^{x,\tau-\frac{1}{2}}\left(\hat{i},\hat{j}+\frac{1}{2},\hat{k}+\frac{1}{2}\right)-\vec{H}^{x,\tau-\frac{1}{2}}\left(\hat{i},\hat{j}-\frac{1}{2},\hat{k}+\frac{1}{2}\right)}{\Delta y}$.

So that equation (3.19) can be expressed in the form of FDTD method as

$$\begin{aligned}\vec{E}^{z,\tau}\left(\hat{i},\hat{j},\hat{k}+\frac{1}{2}\right) &= \frac{1-\dfrac{\sigma\left(\hat{i},\hat{j},\hat{k}+\frac{1}{2}\right)\Delta t}{2\varepsilon_0\varepsilon_r\left(\hat{i},\hat{j},\hat{k}+\frac{1}{2}\right)}}{1+\dfrac{\sigma\left(\hat{i},\hat{j},\hat{k}+\frac{1}{2}\right)\Delta t}{2\varepsilon_0\varepsilon_r\left(\hat{i},\hat{j},\hat{k}+\frac{1}{2}\right)}}\vec{E}^{z,\tau-1}\left(\hat{i},\hat{j},\hat{k}+\frac{1}{2}\right) \\ &+ \frac{\dfrac{\Delta t}{\varepsilon_0\varepsilon_r\left(\hat{i},\hat{j},\hat{k}+\frac{1}{2}\right)}}{1+\dfrac{\sigma\left(\hat{i},\hat{j},\hat{k}+\frac{1}{2}\right)\Delta t}{2\varepsilon_0\varepsilon_r\left(\hat{i},\hat{j},\hat{k}+\frac{1}{2}\right)}} \\ &\left(\frac{\vec{H}^{y,\tau-\frac{1}{2}}\left(\hat{i}+\frac{1}{2},\hat{j},\hat{k}+\frac{1}{2}\right)-\vec{H}^{y,\tau-\frac{1}{2}}\left(\hat{i}-\frac{1}{2},\hat{j},\hat{k}+\frac{1}{2}\right)}{\Delta x}\right. \\ &\left.-\frac{\vec{H}^{x,\tau-\frac{1}{2}}\left(\hat{i},\hat{j}+\frac{1}{2},\hat{k}+\frac{1}{2}\right)-\vec{H}^{x,\tau-\frac{1}{2}}\left(\hat{i},\hat{j}-\frac{1}{2},\hat{k}+\frac{1}{2}\right)}{\Delta y}\right).\end{aligned} \tag{3.20}$$

Based on equation (3.18), we can write $\vec{H}^x$ as

$$\vec{H}^{x,\,\tau+\frac{1}{2}} = \vec{H}^{x,\,\tau-\frac{1}{2}} - \frac{\Delta t}{\mu_0}\left(\frac{\partial \vec{E}^{z,\,\tau}}{\partial y} - \frac{\partial \vec{E}^{y,\,\tau}}{\partial z}\right). \tag{3.21}$$

From figure 3.1(d), $\vec{H}^x$ is located at $\left(\hat{i}, \hat{j} + \frac{1}{2}, \hat{k} + \frac{1}{2}\right)$. Therefore, the differential of $\vec{E}^{z,\,\tau}$ and $\vec{E}^{y,\,\tau}$ in equation (3.21) are $\frac{\vec{E}^{z,\,\tau}\left(\hat{i}, \hat{j}+1, \hat{k}+\frac{1}{2}\right) - \vec{E}^{z,\,\tau}\left(\hat{i}, \hat{j}, \hat{k}+\frac{1}{2}\right)}{\Delta y}$ and $\frac{\vec{E}^{y,\,\tau}\left(\hat{i}, \hat{j}+\frac{1}{2}, \hat{k}+1\right) - \vec{E}^{y,\,\tau}\left(\hat{i}, \hat{j}+\frac{1}{2}, \hat{k}\right)}{\Delta z}$. By substituting them into equation (3.21), $\vec{H}^{x,\,\tau+\frac{1}{2}}$ can be expressed in the form of FDTD method as

$$\begin{aligned}\vec{H}^{x,\,\tau+\frac{1}{2}}\left(\hat{i}, \hat{j} + \tfrac{1}{2}, \hat{k} + \tfrac{1}{2}\right) &= \vec{H}^{x,\,\tau-\frac{1}{2}}\left(\hat{i}, \hat{j} + \tfrac{1}{2}, \hat{k} + \tfrac{1}{2}\right) - \frac{\Delta t}{\mu_0\left(\hat{i}, \hat{j} + \frac{1}{2}, \hat{k} + \frac{1}{2}\right)} \\ &\left(\frac{\vec{E}^{z,\,\tau}\left(\hat{i}, \hat{j} + 1, \hat{k} + \frac{1}{2}\right) - \vec{E}^{z,\,\tau}\left(\hat{i}, \hat{j}, \hat{k} + \frac{1}{2}\right)}{\Delta y}\right. \\ &\left. + \frac{\vec{E}^{y,\,\tau}\left(\hat{i}, \hat{j} + \frac{1}{2}, \hat{k} + 1\right) - \vec{E}^{y,\,\tau}\left(\hat{i}, \hat{j} + \frac{1}{2}, \hat{k}\right)}{\Delta z}\right).\end{aligned} \tag{3.22}$$

Similarly, other field components can be derived as

$$\begin{aligned}\vec{E}^{x,\,\tau}\left(\hat{i} + \tfrac{1}{2}, \hat{j}, \hat{k}\right) &= \frac{1 - \frac{\sigma\left(\hat{i}+\frac{1}{2}, \hat{j}, \hat{k}\right)\Delta t}{2\varepsilon_0\varepsilon_r\left(\hat{i}+\frac{1}{2}, \hat{j}, \hat{k}\right)}}{1 + \frac{\sigma\left(\hat{i}+\frac{1}{2}, \hat{j}, \hat{k}\right)\Delta t}{2\varepsilon_0\varepsilon_r\left(\hat{i}+\frac{1}{2}, \hat{j}, \hat{k}\right)}} \vec{E}^{x,\,\tau-1}\left(\hat{i} + \tfrac{1}{2}, \hat{j}, \hat{k}\right) \\ &+ \frac{\frac{\Delta t}{\varepsilon_0\varepsilon_r\left(\hat{i} + \frac{1}{2}, \hat{j}, \hat{k}\right)}}{1 + \frac{\sigma\left(\hat{i} + \frac{1}{2}, \hat{j}, \hat{k}\right)\Delta t}{2\varepsilon_0\varepsilon_r\left(\hat{i} + \frac{1}{2}, \hat{j}, \hat{k}\right)}} \\ &\left(\frac{\vec{H}^{z,\,\tau-\frac{1}{2}}\left(\hat{i} + \frac{1}{2}, \hat{j} + \frac{1}{2}, \hat{k}\right) - \vec{H}^{z,\,\tau-\frac{1}{2}}\left(\hat{i} + \frac{1}{2}, \hat{j} - \frac{1}{2}, \hat{k}\right)}{\Delta y}\right. \\ &\left. - \frac{\vec{H}^{y,\,\tau-\frac{1}{2}}\left(\hat{i} + \frac{1}{2}, \hat{j}, \hat{k} + \frac{1}{2}\right) - \vec{H}^{y,\,\tau-\frac{1}{2}}\left(\hat{i} + \frac{1}{2}, \hat{j}, \hat{k} - \frac{1}{2}\right)}{\Delta z}\right),\end{aligned} \tag{3.23}$$

$$\vec{E}^{y,\tau}\left(\hat{i},\hat{j}+\tfrac{1}{2},\hat{k}\right)=\frac{1-\dfrac{\sigma\left(\hat{i},\hat{j}+\frac{1}{2},\hat{k}\right)\Delta t}{2\varepsilon_0\varepsilon_r\left(\hat{i},\hat{j}+\frac{1}{2},\hat{k}\right)}}{1+\dfrac{\sigma\left(\hat{i},\hat{j}+\frac{1}{2},\hat{k}\right)\Delta t}{2\varepsilon_0\varepsilon_r\left(\hat{i},\hat{j}+\frac{1}{2},\hat{k}\right)}}\vec{E}^{x,\tau-1}\left(\hat{i},\hat{j}+\tfrac{1}{2},\hat{k}\right)+\frac{\dfrac{\Delta t}{\varepsilon_0\varepsilon_r\left(\hat{i},\hat{j}+\frac{1}{2},\hat{k}\right)}}{1+\dfrac{\sigma\left(\hat{i},\hat{j}+\frac{1}{2},\hat{k}\right)\Delta t}{2\varepsilon_0\varepsilon_r\left(\hat{i},\hat{j}+\frac{1}{2},\hat{k}\right)}}\left(\frac{\vec{H}^{x,\tau-\frac{1}{2}}\left(\hat{i},\hat{j}+\frac{1}{2},\hat{k}+\frac{1}{2}\right)-\vec{H}^{z,\tau-\frac{1}{2}}\left(\hat{i},\hat{j}+\frac{1}{2},\hat{k}-\frac{1}{2}\right)}{\Delta z}-\frac{\vec{H}^{z,\tau-\frac{1}{2}}\left(\hat{i}+\frac{1}{2},\hat{j}+\frac{1}{2},\hat{k}\right)-\vec{H}^{z,\tau-\frac{1}{2}}\left(\hat{i}-\frac{1}{2},\hat{j}+\frac{1}{2},\hat{k}\right)}{\Delta x}\right), \tag{3.24}$$

$$\vec{H}^{y,\tau+\frac{1}{2}}\left(\hat{i}+\tfrac{1}{2},\hat{j},\hat{k}+\tfrac{1}{2}\right)=\vec{H}^{y,\tau-\frac{1}{2}}\left(\hat{i}+\tfrac{1}{2},\hat{j},\hat{k}+\tfrac{1}{2}\right)-\frac{\Delta t}{\mu_0\left(\hat{i}+\frac{1}{2},\hat{j},\hat{k}+\frac{1}{2}\right)}\left(\frac{\vec{E}^{x,\tau}\left(\hat{i}+\frac{1}{2},\hat{j},\hat{k}+1\right)-\vec{E}^{x,\tau}\left(\hat{i}+\frac{1}{2},\hat{j},\hat{k}\right)}{\Delta z}+\frac{\vec{E}^{z,\tau}\left(\hat{i}+1,\hat{j},\hat{k}+\frac{1}{2}\right)-\vec{E}^{z,\tau}\left(\hat{i},\hat{j},\hat{k}+\frac{1}{2}\right)}{\Delta x}\right), \tag{3.25}$$

$$\vec{H}^{z,\tau+\frac{1}{2}}\left(\hat{i}+\tfrac{1}{2},\hat{j}+\tfrac{1}{2},\hat{k}\right)=\vec{H}^{z,\tau-\frac{1}{2}}\left(\hat{i}+\tfrac{1}{2},\hat{j}+\tfrac{1}{2},\hat{k}\right)-\frac{\Delta t}{\mu_0\left(\hat{i}+\frac{1}{2},\hat{j}+\frac{1}{2},\hat{k}\right)}\left(\frac{\vec{E}^{y,\tau}\left(\hat{i}+1,\hat{j}+\frac{1}{2},\hat{k}\right)-\vec{E}^{y,\tau}\left(\hat{i},\hat{j}+\frac{1}{2},\hat{k}\right)}{\Delta x}+\frac{\vec{E}^{x,\tau}\left(\hat{i}+\frac{1}{2},\hat{j}+1,\hat{k}\right)-\vec{E}^{x,\tau}\left(\hat{i}+\frac{1}{2},\hat{j},\hat{k}\right)}{\Delta y}\right). \tag{3.26}$$

3.2.4 Computational resources

Computational time and memory usages of a FDTD calculation are influenced by both the used grid size and the selected time interval. A more accurate result can be obtained by using a finer grid size in computation. However, it is not always favorable for us to shrink the grid size in FDTD calculations, there is something else to consider other than accuracy, such as memory usage and computational time. In other words, the decent grid size must be used to fulfill accuracy, memory and computational time requirements. In general, the used grid size will be $\frac{\lambda}{10} \leqslant \Delta x \leqslant \frac{\lambda}{4}$ per Nyquist stability criterion [10]. According to the Courant–Friedrichs–Lewy condition [11], the used time interval Δt must satisfy the below condition to achieve stable results

$$v\Delta t \leqslant \frac{1}{\sqrt{\left(\frac{1}{\Delta x}\right)^2 + \left(\frac{1}{\Delta y}\right)^2 + \left(\frac{1}{\Delta z}\right)^2}}, \tag{3.27}$$

where v is the wave speed. An example to demonstrate the differences between stable and instable FDTD calculations can be seen in figure 3.3. Assuming we have a two dimensional free space domain with 200 × 200 nm^2 dimension, as seen in figure 3.3(a), where $\Delta x = \Delta y = 1$ nm, which $< \frac{\lambda}{4}$, thus we have enough space resolution. A Gaussian-shaped EUV source (s-polarized) with 13.5 nm CW and 0.8 nm bandwidth (BW) is excited at the center of the computational domain. In this case we apply the most used FDTD boundary condition, which is ABC to the calculations, and let the source propagate for 30 time steps. Thus, the wavefront can be propagated outwardly without touching the boundaries. According to equation (3.27), the numerical instability arises if the used time interval $\Delta t > 1.92$ s. The marginal stable case can be seen in figure 3.3(b). The used time interval of figure 3.3(b) is $\Delta t = 1.92$ s, and the $|\vec{E}^z|$ is plotted at the moment of $30\Delta t$. From figure 3.3(b) readers can see the Gaussian-shaped wave is steadily propagated outward from the domain center. Conversely, $\Delta t > 19.2$ s is used in figure 3.3(c) which should cause numerical instability. As a result, the $|\vec{E}^z|$ jumps up to 3.6×10^{56} at the moment of $25\Delta t$ which violates the energy conservation principle. After verifying the both space resolution and numerical stability, the next step is to estimate the memory budget.

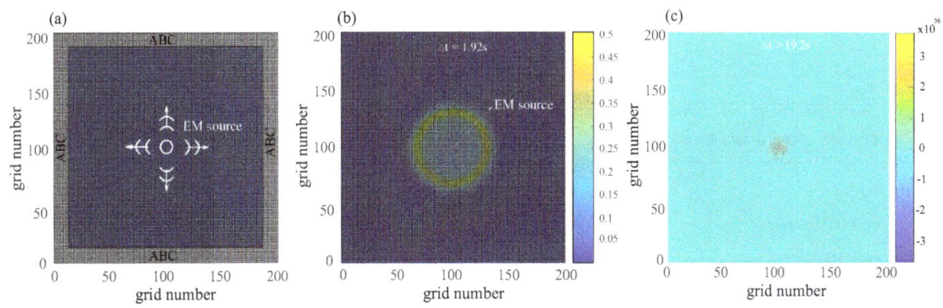

Figure 3.3. FDTD calculation cases, in which (a) computational layout and source setup, (b) stable results at $30\Delta t$ and (c) instable results at $25\Delta t$.

If there are N_c cells used in FDTD computation with double precision, the memory required in a single time interval is

$$N_c n_e \times 8 \text{ bytes}, \tag{3.28}$$

where n_e is the number of unknowns in a cell, for example, the electric fields, the magnetic fields, the current densities, etc. The computational time is estimated by referring to the mesh numbers as

$$t_{\text{comp}} = \xi \left(\frac{\Delta_{\text{comp}}}{\Delta_{\text{ref}}} \right)^{n_{\text{d}}+1} t_{\text{ref}}, \tag{3.29}$$

where t_{comp} is the computational time, ξ is a factor related to the CPU's floating point operations per second (FLOPs), Δ_{comp} is the grid numbers along one axis in the current computational domain, Δ_{ref} is the referenced grid numbers along one axis, n_{d} is the number of dimensions applied in the computational domain and t_{ref} is the reference computational time of specific grid numbers. As mentioned in equation (3.27), to satisfy the stability requirement, the selection of time interval has a lot to do with the space domain grid size. Thus, the order of power is 4 in equation (3.29) since there are three dimensions in the space domain and one dimension in the time domain.

Table 3.1 demonstrates the memory and computational time usage of four different grid numbers used in FDTD calculations. 2D cases were used in table 3.1, the total computational domain is 250×600 nm^2 which is applied to all four cases. However, different mesh resolutions are applied to these four cases. In the computational domain we created a 40 layers Mo/Si based ML structure, an EUV light with 13.5 nm CW and 0.8 nm BW illuminated toward the ML normally. The grid size of case one is 0.675 (i.e., $(1/20)\lambda$), for case two is 0.2085 nm (i.e., Si thickness $\times$ 1/20), for case three is 0.139 nm (i.e., Mo thickness $\times$ 1/20) and for case four is 0.0695 nm (i.e., Mo thickness $\times$ 1/40). There will be an inevitable increment of total grid numbers when applying a finer grid size. All calculations were finished on a hexa core 2.60 GHz Intel i7-9750H CPU with 8GB DDR4 RAM. The calculations were truncated when the total energy in the computational domain drops to 10^{-4} of the input energy. The memory usage can be estimated by using equation (3.28) with $n_e = 14$ in our calculations, these estimated values are compared with the actual memory usage in table 3.1. As for the computational time, the case of minimum grid number (3.2×10^5) is used as the reference to estimate the computational time

Table 3.1. Comparisons of memory and computational time usage between theoretical estimations and numerical results from FDTD calculations for 2D cases.

Grid numbers	3.2×10^5	3.4×10^6	7.7×10^6	3.1×10^7
Memory (execution)	42 MB	352 MB	791 MB	3.1 GB
Memory (estimation)	38 MB	360 MB	810 MB	3.2 GB
Run time (sec, execution)	46	1140	4850	42 659
Run time (sec, estimation)	46	1560	5267	42 142

consumption for other cases. From the results listed in table 3.1, the readers can see good fits between the estimations and the actual calculations.

3.3 Boundary conditions

The FDTD method introduced by Yee distributes a space into staggered Cartesian grids. Before performing the calculation, an important step is to define an enclosed space with boundaries so that we know the required computational resource. To design an EUV mirror, we need to calculate the EUV scattering from the mirror. The ideal scenario is to simulate this kind of scattering problem in an open area, which means that no reflection from the boundaries will interrupt the calculations. However, this requires to build-up infinite space grids and is not possible to be realized in the FDTD scheme. Therefore, it is always assumed that somewhere in the grids is truncated. The below two boundary conditions are common in FDTD calculations and are used in EUV mirror design problems that will be described in chapter 4.

3.3.1 Absorption boundary condition

The FDTD method introduced by Yee does not include the concept of absorption boundary condition (ABC), since ABC cannot be directly obtained from the numerical algorithms of Maxwell's equations defined finite-difference systems. Instead, the default BCs are the perfect electric conductors (PECs), which means by default the boundaries are all ideal lossless metallic surfaces. The default PEC boundaries are not favorable for EUV mirror design, because of the huge ratio of the EUV ML size to the EUV wavelength, the fine meshes must be applied to ensure the accuracy. Therefore, the computational domain should be as small as possible in consideration of computational resources availability. In order to make such a limited space equivalent to unlimited space, the ABC is used to simulate the infinite open area. The 2D illustrations of how to use ABCs to simulate an open space can be seen in figures 3.4(a) and (b). To simulate the source scattering in an open area as illustrated in figure 3.4(a), the ABCs can be used to encompass the source and the scattering object as shown in figure 3.4(b). The ABCs will absorb the scattering waves. An ideal ABC should only be the outward transfer wave rather than inward reflection wave. To make the best use of imitating open area, the distances from the scattering object to all the ABCs must be long enough to prevent artificial coupling. However, this increases the size of the computational domain, which is equal to extending the computational time and to enlarging the memory usage. For instance, assuming we have a cubical scattering object at the center of the computational domain. The total mesh numbers of the computational domain are $20 \times 20 \times 20 = 20^3$. Let's assume the minimum required distance from each surface of the scattering object to the ABC should be 5 grids. By doing this, the entire computational domain is enlarged from $20^3 = 8000$ grids to $25^3 = 15\ 625$ grids. According to equations (3.28) and (3.29), the required memory usage will be doubled and the required computational time will become $\times$ 16 times longer after we enlarge the

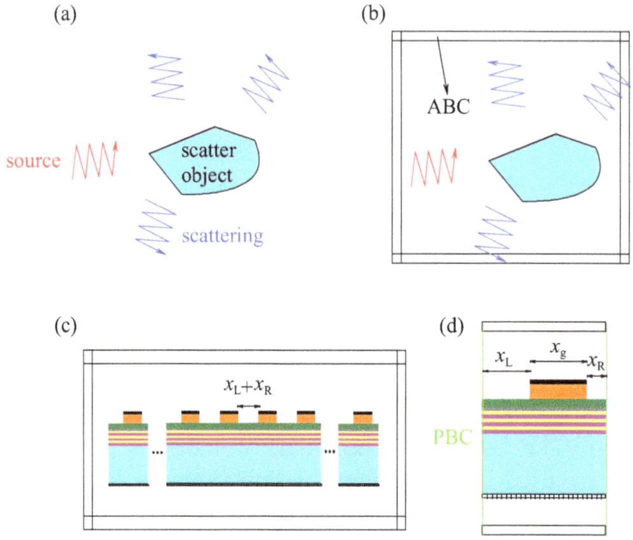

Figure 3.4. (a) A source, a scatter object and resulted wave scatterings in an open space, which is not the case in FDTD simulations. (b) A limited space encompassed by ABCs. This is how the FDTD method simulates an open environment. (c) An ML stack based mirror deposited by periodic gratings encompassed by ABCs. Owing to the long periodicity of the gratings, this will consume a lot of computational resources in FDTD calculation. (d) A limited space that is constructed by PBCs at the left and the right, while the up and the bottom boundaries are ABCs. This way the computational domain is dramatically reduced compared to (c).

Table 3.2. A list of representative ABCs in FDTD calculation.

Differential-based ABC	Mur ABC [12]
	Higdon ABC [13]
	Liao ABC [14]
	Lindman ABC [15]
Material-based ABC	Rappaport PML [16]
	Berenger PML [17]
	Sacks PML [18]
	Convolutional PML [14]
	Uniaxial PML [14]

computational domain. There are various ABCs that were published as listed in table 3.2, they can be grouped into two categories:

- **Differential-based ABC**: This category is based on the approximation of the one-way wave equations through factorization of wave equations. It generates result with lower accuracy, and basically induces no extra features such as memory consumption.
- **Material-based ABC**: This kind of ABC is also called perfect matched layer (PML), to induct absorption material outside the truncation boundary, so the

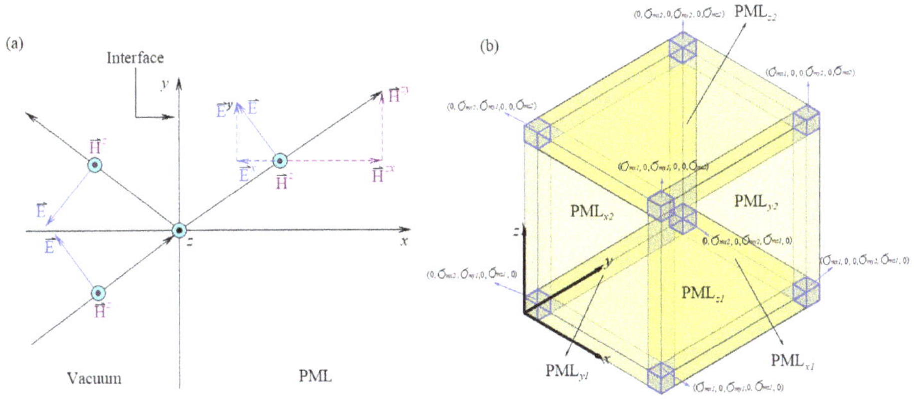

Figure 3.5. (a) An *s*-polarized wave propagates from vacuum to PML. (b) The PML's implementation in 3D.

EM wave will decay through it. It generates result with higher accuracy. However, extra memory and computational time requirements are induced.

Among all types of ABCs, Berenger's PML is the most effective ABC among them. Thus, Berenger's PML is chosen for our EUV mirror design. To understand how Berenger's PML works, we take a 2D illustration as an example as shown in figure 3.5(a) to describe the principle of the PML. Assuming we have a *s*-polarized plane wave propagating from -*x* to +*x* along the *x*-direction, which is from the vacuum toward the PML (the PML boundary starts somewhere along +*x*). The impedance of the vacuum is $Z_0 = \sqrt{\frac{\mu_0}{\varepsilon_0}}$ and the impedance of the PML is $Z_{\rm PML} = \sqrt{\frac{\mu PML + \frac{\sigma m}{j\omega}}{\varepsilon PML + \frac{\sigma e}{j\omega}}}$, where σ_e and σ_m stand for the electric and magnetic conductivities of the PML. The Berenger's PML obeys $Z_0 = Z_{\rm PML}$, therefore, no reflection occurs because the reflection coefficient = 0. Since there will be no reflection from the PML boundary, the EM wave is regarded as totally passing through the PML. Thus, the *x* components of the *s*-polarized wave in the PML can be described as

$$\mu_0 \frac{\partial \vec{H}^{zx}}{\partial t} + \sigma_m \vec{H}^{zx} = -\frac{\partial \vec{E}^{y}}{\partial x} \tag{3.30}$$

and

$$\varepsilon_0 \frac{\partial \vec{E}^{y}}{\partial t} + \sigma_e \vec{E}^{y} = -\frac{\partial \vec{H}^{zx}}{\partial x} \tag{3.31}$$

where $\vec{H}^{zx}$ stands for the *x* component of $\vec{H}^{z}$. The *y* components of the plane wave can be described as

$$\mu_0 \frac{\partial \vec{H}^{zy}}{\partial t} = \frac{\partial \vec{E}^{x}}{\partial y} \tag{3.32}$$

and

$$\varepsilon_0 \frac{\partial \vec{E}^x}{\partial t} = \frac{\partial \vec{H}^{zy}}{\partial y} \tag{3.33}$$

where $\vec{H}^{zy}$ stands for the y component of $\vec{H}^z$, and $\vec{H}^z = \vec{H}^{zx} + \vec{H}^{zy}$. Considering a 3D computational domain which is a vacuum and is surrounded by PML boundaries, as shown in figure 3.5(b). The Maxwell's equations are solved by the FDTD method within the vacuum area. The waves in PML can be described through the following equations

$$\mu_0 \frac{\partial \vec{H}^{xy}}{\partial t} + \sigma_{my} \vec{H}^{xy} = -\frac{\partial \vec{E}^z}{\partial y}, \tag{3.34}$$

$$\mu_0 \frac{\partial \vec{H}^{xz}}{\partial t} + \sigma_{mz} \vec{H}^{xz} = \frac{\partial \vec{E}^y}{\partial z}, \tag{3.35}$$

$$\mu_0 \frac{\partial \vec{H}^{yz}}{\partial t} + \sigma_{mz} \vec{H}^{yz} = -\frac{\partial \vec{E}^x}{\partial z}, \tag{3.36}$$

$$\mu_0 \frac{\partial \vec{H}^{yx}}{\partial t} + \sigma_{mx} \vec{H}^{yx} = \frac{\partial \vec{E}^z}{\partial x}, \tag{3.37}$$

$$\mu_0 \frac{\partial \vec{H}^{zx}}{\partial t} + \sigma_{mx} \vec{H}^{zx} = -\frac{\partial \vec{E}^y}{\partial x}, \tag{3.38}$$

$$\mu_0 \frac{\partial \vec{H}^{zy}}{\partial t} + \sigma_{my} \vec{H}^{zy} = \frac{\partial \vec{E}^x}{\partial y}, \tag{3.39}$$

$$\varepsilon_0 \frac{\partial \vec{E}^{xy}}{\partial t} + \sigma_{ey} \vec{E}^{xy} = \frac{\partial \vec{H}^x}{\partial y}, \tag{3.40}$$

$$\varepsilon_0 \frac{\partial \vec{E}^{xz}}{\partial t} + \sigma_{ez} \vec{E}^{xz} = -\frac{\partial \vec{H}^y}{\partial z}, \tag{3.41}$$

$$\varepsilon_0 \frac{\partial \vec{E}^{yz}}{\partial t} + \sigma_{ez} \vec{E}^{yz} = \frac{\partial \vec{H}^x}{\partial z}, \tag{3.42}$$

$$\varepsilon_0 \frac{\partial \vec{E}^{yx}}{\partial t} + \sigma_{ex} \vec{E}^{yx} = -\frac{\partial \vec{H}^z}{\partial x}, \tag{3.43}$$

$$\varepsilon_0 \frac{\partial \vec{E}^{zx}}{\partial t} + \sigma_{ex} \vec{E}^{zx} = \frac{\partial \vec{H}^y}{\partial x}, \tag{3.44}$$

and

$$\varepsilon_0 \frac{\partial \vec{E}^{zy}}{\partial t} + \sigma_{ey} \vec{E}^{zy} = -\frac{\partial \vec{H}^{x}}{\partial y} \tag{3.45}$$

where σ_{mx}, σ_{my} and σ_{mz} are magnetic conductivities of the PML in x, y and z directions. σ_{ex}, σ_{ey} and σ_{ez} are electric conductivities of the PML in x, y and z directions. $\vec{H}^{xy}$and $\vec{H}^{xz}$ are y and z component of $\vec{H}^{x}$. $\vec{H}^{yz}$ and $\vec{H}^{yx}$ are z and x component of $\vec{H}^{y}$. $\vec{E}^{xy}$and $\vec{E}^{xz}$ are y and z component of $\vec{E}^{x}$. $\vec{E}^{yz}$ and $\vec{E}^{yx}$ are z and x component of $\vec{E}^{y}$. $\vec{E}^{zx}$and $\vec{E}^{zy}$ are x and y component of $\vec{E}^{z}$. Outside the PML boundaries, the PECs are bordered to delimit the entire domain. At both interfaces vacuum-PML$_{x1}$ and vacuum-PML$_{x2}$ normal to the x-axis, the reflection factor does not exist. Thus, outgoing waves can penetrate through PML$_{x1}$ and PML$_{x2}$ (σ_{mx1}, σ_{mx2}, 0, 0, 0, 0) without reflection. In the same way, the outgoing waves can penetrate through PML$_{y1}$, PML$_{y2}$ (0, 0, σ_{my1}, σ_{my2}, 0, 0), PML$_{z1}$ and PML$_{z2}$ (0, 0, 0, 0, σ_{mz1}, σ_{mz2}) without reflection. At the eight corners of the domain, the materials are the PMLs having conductivities equal to those of the adjacent (σ_{mx1}, σ_{mx2}, 0, 0, 0, 0), (0, 0, σ_{my1}, σ_{my2}, 0, 0) and (0, 0, 0, 0, σ_{mz1}, σ_{mz2}). To sum up, there is no reflection at the interfaces lying between the PML surfaces and the corners.

2D demonstrations of PEC and PML for FDTD calculations can be seen in figure 3.6, in which we setup a domain with 200 × 200 nm^2. A s-polarized point source with Gaussian distribution of CW/BW = 13.5 nm/0.8 nm is generated at the center of the domain. In addition, a circular scatter object (an artificial material with $\tilde{n}$ = 0.0797 $-j$ 0.2368) with its radius = 25 nm is located below the point source. In figures 3.6(a) and (b), readers can see only the PEC boundary condition is used. Anyhow, the results from FDTD calculations in these two sub-figures will be the same no matter if the PML boundary condition is used or not. Since no EM wave touches the boundaries, no PML boundary condition needs to be considered. In figure 3.6(a), at the moment of 2.4 × 10^{-17} s (10Δt), the EUV Gaussian source keeps propagating toward all directions. By applying the same EUV source, the | $\vec{E}^{z}$ | distribution with a circular scatter object at 1.9 × 10^{-16} s (80Δt) can be seen, figure 3.6(b). It is clear that not only reflections but also transmissions can be observed nearby the scatter object. The results of figures 3.6(c)–(f) are | $\vec{E}^{z}$ | distributions without scatter object, where figures 3.6(c) and (e) use default PEC boundary condition but figures 3.6(d) and (f) use the PML boundary condition. The |$\vec{E}^{z}$| distribution at 3.8 × 10^{-16} s (160Δt) are shown in figures 3.6(c) and (d). If we compare the |$\vec{E}^{z}$| between these two figures, stronger |$\vec{E}^{z}$| inside the point source and more energy on the wavefront can be observed in figure 3.6(c) because of the reflections from PEC boundaries. In contrast, in figure 3.6(d) the energy inside the Gaussian source stays at the relatively lower level and no wavefront energy has been intensified. The obvious differences between the PEC and PML boundary are shown in figures 3.6(e) and (f), which are the | $\vec{E}^{z}$ | distributions at 4 × 10^{-16} s (170Δt) for PEC and PML boundary implementations, respectively. As seen from figure 3.6(e), the severe reflections from PEC boundaries are observed. The reflections bring more | $\vec{E}^{z}$ | energies both inside the Gaussian source and on the wavefront than in figure 3.6(c). For figure 3.6(f), the PML boundaries

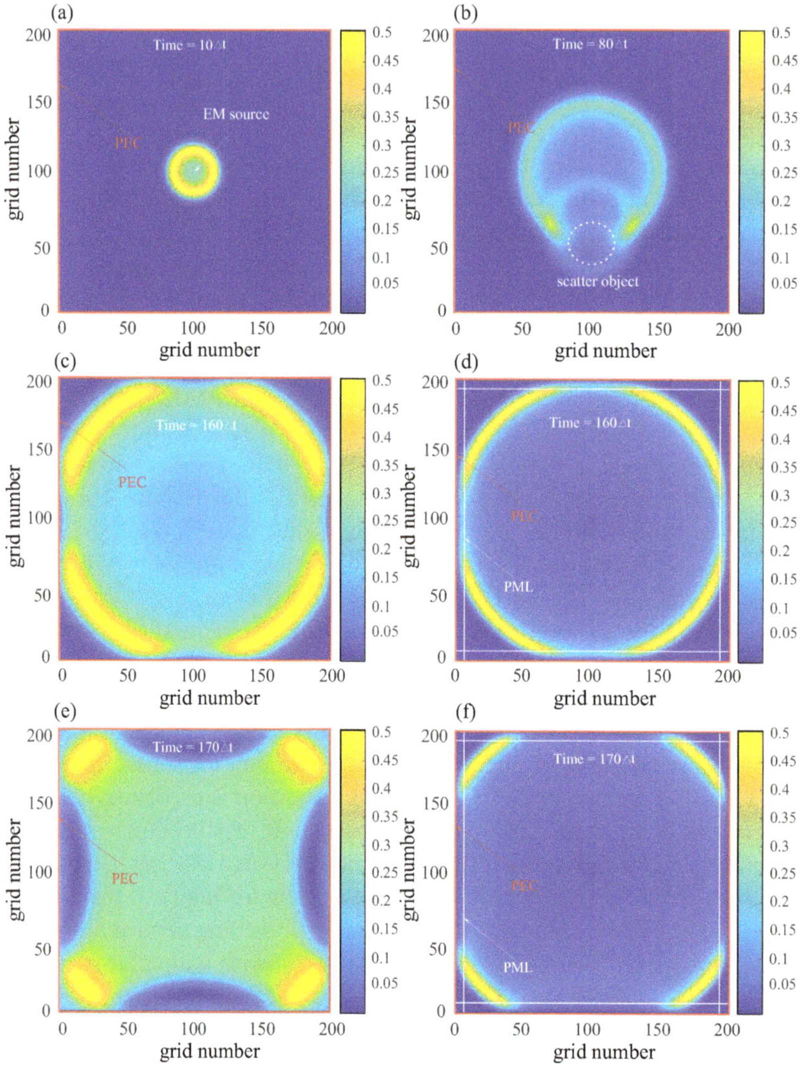

Figure 3.6. $|\vec{E}^z|$ distributions of different 2D FDTD simulation cases. (a) A Gaussian source spreads in vacuum at 2.4×10^{-17} s. (b) The EM scattering and transmission after the Gaussian source hits the scatter object at 1.9×10^{-16} s. (c) and (e) are the $|\vec{E}^z|$ distributions in vacuum by applying PEC boundaries at 3.8×10^{-16} s and 4×10^{-16} s. (d) and (f) are the $|\vec{E}^z|$ distributions in vacuum by applying PML boundaries at 3.8×10^{-16} s and 4×10^{-16} s.

absorb energy well, thus the wavefront disappears when reaching the boundaries. This is how PML is used to simulate the open space in the FDTD calculation.

3.3.2 Periodic boundary condition (PBC)

The periodic boundary condition is a kind of boundary condition which can be used to simulate a huge or infinite object within a limited computational domain. PBC is required in EUV ML mirror design owing to the below reasons:

- The scale of an EUV mirror is the inch (the common dimensions of a reticle are 6×6 inch^2 with a thickness of 0.25 inch [19]) while the scale of the EUV light is the nanometer (13.5 nm for EUV lithography). With such a huge scale difference, to design an EUV mirror by using the FDTD method with PML boundary conditions becomes a large scale problem.
- Periodic features such as periodic contact holes are common study cases for an EUV mirror. PBC can be used to simulate infinite periodic features on or within the EUV mirror without memory shortage concern.

The effect of PBC is illustrated in figures 3.4(c) and (d), assuming there is a 2D design case that is a ML mirror deposited by periodic gratings. Normally the scale of the grating pitch < 1 μm, thus there will be huge amount of gratings (over 1 52 400 pitches for a 6 inch mirror). With ABCs bounded computational domain as shown in figure 3.4(c), if the grid size = 0.675 nm is applied. The approximate grid numbers to mesh the mirror will be 10 160 000 × 243 840 000, which requires 2 307 272 GB to solve this problem (with $n_e = 1$). In contrast, by applying PBC at the left and the right sides, the computational domain is dramatically reduced, as shown in figure 3.4(d). The PBCs facilitates the FDTD method to calculate the results of infinite periods and shows the results for only one period. In addition, to make the best use of ABC in figure 3.4(c), there are required spaces from the edge of the grating to the left and the right ABCs. These spaces consume additional computational resources which are not an issue anymore in figure 3.4(d). By applying the same grids size = 0.675 nm, the grid numbers to mesh the one-period mirror with 1 μm pitch are 10 160 000 × 1481, which requires only 15 GB for the FDTD method to solve the problem. To implement PBC boundaries in 3D, consider a computational domain of size $\psi_x \times \psi_y \times \psi_z$ discretized into $N_x \times N_y \times N_z$ grids. Likewise to figure 3.1(d), each Yee cell is bounded by tangential $\vec{E}$ fields and normal $\vec{H}$ fields. As illustrated in figure 3.7, taking two y–z PBC boundaries as an example to demonstrate how the fields on these two planes are updated between them. In the figure readers can see all the $\vec{H}$ fields require updates from the $\vec{E}$ fields, but the boundary $\vec{E}$ fields require $\vec{H}$ fields from outside the computational domain to be updated. The external supplementary $\vec{H}$ nodes can be updated by shifting $\vec{H}$ fields from the prior period, which fall within the computational domain. The $\vec{E}$ fields residing on the remaining sides can be computed by directly shifting the respective opposite planes, including the fields on the boundaries between the updated planes and the non-updated planes. If the planes at $i = N_x + ½, j = N_y + ½$ and $k = N_z + ½$ are the places to have the external supplementary $\vec{H}$ nodes. The $\vec{H}$ fields of a 3D PBC boundary can be described as

$$\vec{H}^{x,\,\tau-\frac{1}{2}}\left(\hat{i}, \hat{j} + N_y + \frac{1}{2}, \hat{k} + \frac{1}{2}\right) = \vec{H}^{x,\,\tau-\frac{1}{2}}\left(\hat{i}, \hat{j} + \frac{1}{2}, \hat{k} + \frac{1}{2}\right) e^{-jk_y\Psi_y}, \tag{3.46}$$

$$\vec{H}^{x,\,\tau-\frac{1}{2}}\left(\hat{i}, \hat{j} + \frac{1}{2}, \hat{k} + N_z + \frac{1}{2}\right) = \vec{H}^{x,\,\tau-\frac{1}{2}}\left(\hat{i}, \hat{j} + \frac{1}{2}, \hat{k} + \frac{1}{2}\right) e^{-jk_z\Psi_z}, \tag{3.47}$$

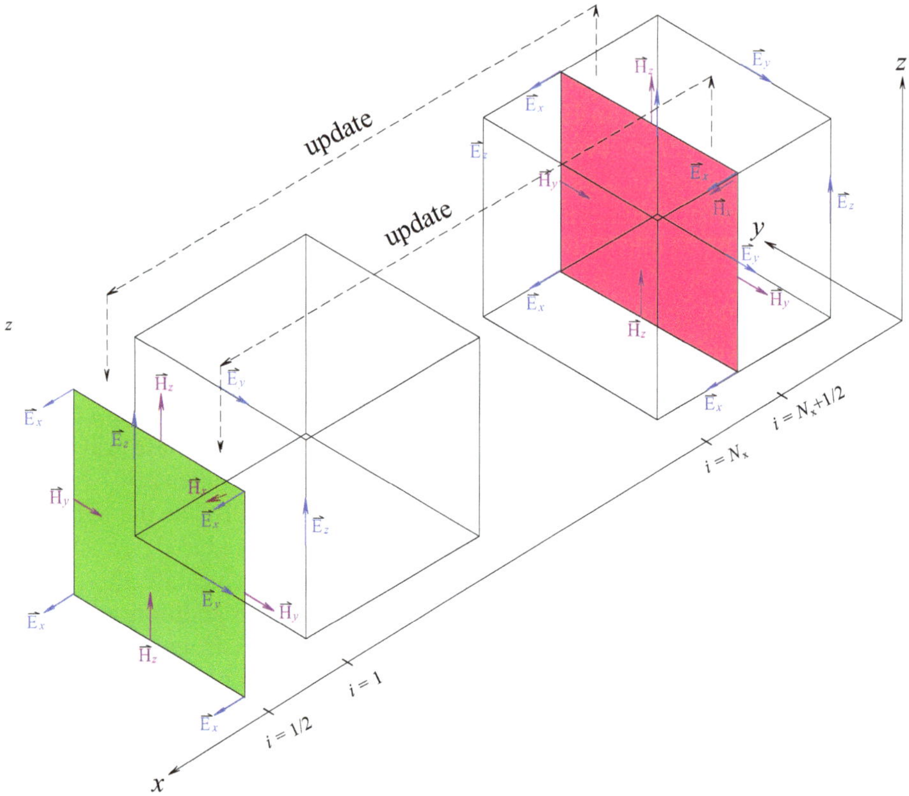

Figure 3.7. Periodic boundary condition implementing principle at single time interval in 3D. The $\vec{H}$ and $\vec{E}$ fields on the green plane are updated by the fields on the supplementary plan (pink plane).

$$\vec{H}^{y,\,\tau-\frac{1}{2}}\left(\hat{i} + N_x + \tfrac{1}{2}, \hat{j}, \hat{k} + \tfrac{1}{2}\right) = \vec{H}^{y,\,\tau-\frac{1}{2}}\left(\hat{i} + \tfrac{1}{2}, \hat{j}, \hat{k} + \tfrac{1}{2}\right)e^{-jk_x\Psi_x}, \tag{3.48}$$

$$\vec{H}^{y,\,\tau-\frac{1}{2}}\left(\hat{i} + \tfrac{1}{2}, \hat{j}, \hat{k} + N_z + \tfrac{1}{2}\right) = \vec{H}^{y,\,\tau-\frac{1}{2}}\left(\hat{i} + \tfrac{1}{2}, \hat{j}, \hat{k} + \tfrac{1}{2}\right)e^{-jk_z\Psi_z}, \tag{3.49}$$

$$\vec{H}^{z,\,\tau-\frac{1}{2}}\left(\hat{i} + N_x + \tfrac{1}{2}, \hat{j} + \tfrac{1}{2}, \hat{k}\right) = \vec{H}^{z,\,\tau-\frac{1}{2}}\left(\hat{i} + \tfrac{1}{2}, \hat{j} + \tfrac{1}{2}, \hat{k}\right)e^{-jk_x\Psi_x} \tag{3.50}$$

and

$$\vec{H}^{z,\,\tau-\frac{1}{2}}\left(\hat{i} + \tfrac{1}{2}, \hat{j} + N_y + \tfrac{1}{2}, \hat{k}\right) = \vec{H}^{z,\,\tau-\frac{1}{2}}\left(\hat{i} + \tfrac{1}{2}, \hat{j} + \tfrac{1}{2}, \hat{k}\right)e^{-jk_y\Psi_y}. \tag{3.51}$$

The tangential $\vec{E}$ fields on the planes $i = 0$, $j = 0$ and $k = 0$ can subsequently be updated as below

$$\vec{E}^{x,\,\tau}\left(\hat{i} + \tfrac{1}{2}, \hat{j}, \hat{k}\right) = \vec{E}^{x,\,\tau}\left(\hat{i} + \tfrac{1}{2}, \hat{j} + N_y, \hat{k}\right)e^{jk_y\Psi_y}, \tag{3.52}$$

$$\vec{E}^{x,\tau}\left(\hat{i}+\tfrac{1}{2},\hat{j},\hat{k}\right)=\vec{E}^{x,\tau}\left(\hat{i}+\tfrac{1}{2},\hat{j},\hat{k}+N_z\right)e^{jk_z\Psi_z}, \tag{3.53}$$

$$\vec{E}^{y,\tau}\left(\hat{i},\hat{j}+\tfrac{1}{2},\hat{k}\right)=\vec{E}^{y,\tau}\left(\hat{i},\hat{j}+\tfrac{1}{2},\hat{k}+N_z\right)e^{jk_z\Psi_z}, \tag{3.54}$$

$$\vec{E}^{y,\tau}\left(\hat{i},\hat{j}+\tfrac{1}{2},\hat{k}\right)=\vec{E}^{y,\tau}\left(\hat{i}+N_x,\hat{j}+\tfrac{1}{2},\hat{k}\right)e^{jk_x\Psi_x}, \tag{3.55}$$

$$\vec{E}^{z,\tau}\left(\hat{i},\hat{j},\hat{k}+\tfrac{1}{2}\right)=\vec{E}^{z,\tau}\left(\hat{i}+N_x,\hat{j},\hat{k}+\tfrac{1}{2}\right)e^{jk_x\Psi_x} \tag{3.56}$$

and

$$\vec{E}^{z,\tau}\left(\hat{i},\hat{j},\hat{k}+\tfrac{1}{2}\right)=\vec{E}^{z,\tau}\left(\hat{i},\hat{j}+N_y,\hat{k}+\tfrac{1}{2}\right)e^{jk_y\Psi_y}. \tag{3.57}$$

Moreover, the $\vec{H}$ fields on the planes $i = 0$, $j = 0$ and $k = 0$ may also be used to update the $\vec{H}$ fields on the opposite planes.

To sum up, the FDTD PBC construction follows the below steps (for each time interval):

- At time step $\tau = n - ½$, update all $\vec{H}$ fields (assume we have n time steps in total), except for the supplementary $\vec{H}$ nodes at $i = N_x + ½$, $j = N_y + ½$ and $k = N_z + ½$.
- Update the supplementary $\vec{H}$ nodes by shifting $\vec{H}$ fields from $i = ½$, $j = ½$ and $k = ½$.
- At time step $\tau = n$, update all $\vec{E}$ fields, except for those along $i = 0$, $j = 0$ and $k = 0$.
- Update the $\vec{E}$ fields along $i = 0$, $j = 0$ and $k = 0$ by shifting the fields on the opposite planes.

3.4 Efficient FDTD method—the equivalent layer approach

The memory requirement and the computational time usage of the FDTD computation mainly depend on the grid numbers used in the computational domain, which directly link to the defined grid size, the used wavelength and the structure complexity of the scattering object. It has been described in section 3.2.4, more accurate results can be obtained by using finer meshes. However, the price we need to pay is to spend a longer computational time and also to prepare the huge amount of memory during computation [8]. As we already introduced in chapter 2, the ML structure of the EUV mirror contains multiple thin films, that are deposited by Mo and Si periodically (sometimes also mixed with another few subsidiary layers [20, 21]). Owing to the thin thickness of each film in ML (less than 5 nm) and numerous interfaces between layers, the meshes discretized for FDTD computation require more careful consideration. To eliminate this concern, the equivalent layer approach is applied to the ML structure to improve the efficiency of the FDTD calculation. This way, the ML structure is treated as an equivalent bulk layer with an equivalent refractive index throughout FDTD calculation. The advantage to apply this equivalent layer for EUV mirror design is that we avoid the meshing problem in FDTD calculations, and thus accelerate the calculation speed. This is what we called the efficient FDTD method.

3.4.1 Equivalent impedance

The concepts of equivalent layer have long been studied. For instance, Ono proposed that the crystal structure is identical to the multiple thin-film structure and can be treated as an equivalent layer through the matrix method [22]. Torrungrueng and Lamultree simulated the ML structures and calculated the reflectance by using finite lossless periodically loaded transmission line [23]. Our methodology starts with the analogy between the ML structure and the finite transmission line. As a uniform plane wave travels through a stack of ML structure such as an EUV mirror, the finite transmission line model can be used to analogize the impedance of the ML structure. As shown in figure 3.12, take a three-layer ML structure as an example (i.e., two interfaces with its intrinsic impedance η_1, η_2 and η_3). By obtaining an equivalent intrinsic impedance η_{eq} for the entire ML, by means of transmission lines linking, as indicated in figure 3.8, the three-layer structure is simplified to a single bulk layer. This concept can be applied to the ML with more layers, here we use a three-layer system to derive the formulations. For the normal incidence case, the equivalent intrinsic impedance of layer 2 and layer 3 is expressed as [24]

$$\eta_{2\_3} = \eta_2 \frac{\eta_3 + \eta_2 \tanh \gamma_2 T}{\eta_2 + \eta_3 \tanh \gamma_2 T}, \tag{3.58}$$

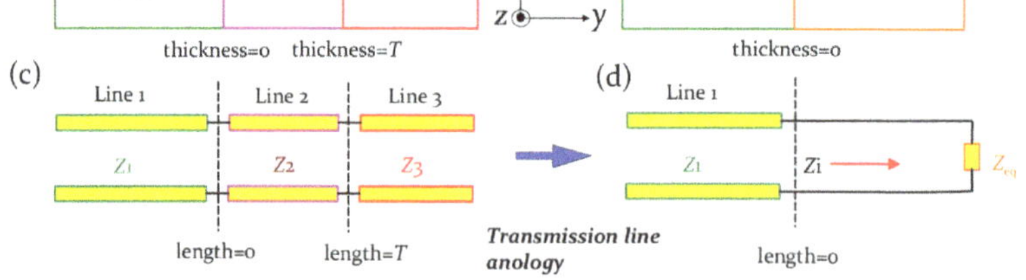

Figure 3.8. Wave propagates through multiple dielectric layers analogous to connected finite transmission lines: (a) a three-layer ML structure, (b) the equivalent one-layer structure, (c) finite transmission lines with different impedances, and (d) the equivalent one-length finite transmission line. In (a) and (b), the symbols η_1, η_2, and η_3 are the intrinsic impedances of layers 1, 2, and 3, η_{eq} is the equivalent intrinsic impedance of the equivalent layer; In (c) and (d), Z_1, Z_2, and Z_3 are the characteristic impedance of transmission lines 1, 2, and 3, and Z_{eq} is its equivalent characteristic impedance.

where T is the thickness of layer 2, γ_2 stands for the characteristic propagation constant, which can be described by using the applied wavelength and the corresponding refractive index of the medium

$$\gamma_2 = j\frac{2\pi\tilde{n}_2}{\lambda}, \tag{3.59}$$

where $\tilde{n}_2$ stands for the complex refractive index of layer 2. For three finite transmission lines liked together with different impedances Z_1, Z_2, and Z_3 as illustrated in figure 3.8(c), given a characteristic propagation constant γ_2 and length T of the second transmission line, the overall input impedance observed from transmission line 1 as shown in figure 3.8(d) is

$$Z_{\text{in}} = Z_2\frac{Z_3 + Z_2\tanh\gamma_2 T}{Z_2 + Z_3\tanh\gamma_2 T}. \tag{3.60}$$

Readers can tell that equations (3.58) and (3.60) are comparable, meaning that the equivalent intrinsic impedance observed from the first ML layer is analogous to the input impedance of the multiple finite-length transmission lines. For glancing wave illuminations, the equivalent intrinsic impedance observed from the first ML layer can be derived from the modified transmission line theory. As shown in figure 3.8(a), as a wave illuminates from different angles Φ_{1i} toward the ML. The equivalent intrinsic impedance for combination of layers 2 and 3 in figure 3.8(b) is

$$\eta_{eq} = \eta_2'\frac{\eta_3' + \eta_2'\tanh\gamma_2 T'}{\eta_2' + \eta_3'\tanh\gamma_2 T'}, \tag{3.61}$$

with the equivalent length $T' = \frac{T}{cos\Phi_{2t}}$. θ_{2t} stands for the angles of transmission in layer 2, which can be expressed as

$$\theta_{2t} = \begin{bmatrix} \theta_{21} \\ \theta_{22} \\ \theta_{23} \\ \vdots \\ \theta_{2x} \end{bmatrix} \tag{3.62}$$

where θ_{2x} implies the xth transmission in layer 2. Based on Fresnel equations, the reflection angles in layer 1 (θ_{1r}) and in layer 2 (θ_{2r}) are equal to the incident angles in layer 1 (θ_{1i}) and layer 2 (θ_{2t}). Thus, we have $\theta_{1i} = \theta_{1r}$ and $\theta_{2t} = \theta_{2r}$. In addition, the intrinsic impedance of layer 2 for s- and p-polarizations are

$$s - \text{polarized}: \quad \eta_2' = \eta_2\cos\theta_{2t}, \tag{3.63}$$

and

$$p - \text{polarized}: \quad \eta_2' = \frac{\eta_2}{\cos\theta_{2t}}. \tag{3.64}$$

For layer 3, the intrinsic impedance for *s*- and *p*-polarizations are

$$s - \text{polarized}: \quad \eta_3' = \eta_3 \cos\theta_{3t}, \tag{3.65}$$

$$p - \text{polarized}: \quad \eta_3' = \frac{\eta_3}{\cos\theta_{3t}} \tag{3.66}$$

where $\theta_{3t} = \begin{bmatrix} \theta_{31} \\ \theta_{32} \\ \theta_{33} \\ \vdots \\ \theta_{3x} \end{bmatrix}$ already are the transmitted angles in layer 3. By applying above equations to equation (3.60), the equivalent impedance looking toward from transmission line 1 in figures 3.8(c) and (d) is

$$Z_{eq} = Z_2' \frac{Z_3' + Z_2' \tanh\gamma_2 T'}{Z_2' + Z_3' \tanh\gamma_2 T'}. \tag{3.67}$$

The impedances of the second transmission line for *s*- and *p*-polarizations are

$$s - \text{polarized}: \quad Z_2' = Z_2 \cos\theta_{2t}, \tag{3.68}$$

and

$$p - \text{polarized}: \quad Z_2' = \frac{Z_2}{\cos\theta_{2t}}. \tag{3.69}$$

For the third transmission line, the impedance for *s*- and *p*-polarizations are

$$s - \text{polarized}: \quad Z_3' = Z_3 \cos\theta_{3t}, \tag{3.70}$$

and

$$p - \text{polarized}: \quad Z_3' = \frac{Z_3}{\cos\theta_{3t}}. \tag{3.71}$$

The equivalent refractive index of the equivalent layer therefore can be described as

$$\tilde{n}_{eq} = \frac{1}{Z_{eq}}. \tag{3.72}$$

3.4.2 Verification of equivalent layer

In this section, we would like to further demonstrate the usability of the equivalent layer mentioned in section 3.4.1. A Fresnel's equations based in-house solver was used to verify the impedance of the equivalent layer. The Fresnel's equations based solver is reliable and can be considered as an analytical result, which has been compared with the public solver from the Center for X-ray Optics (CXRO) [65] as shown in figures 3.9(a) and (b). Figures 3.9(a) and (b) show the reflectivities from $\lambda = 12.8$–14 nm calculated by the Fresnel's equations based solver and CXRO solver under *s*- and *p*-polarizations (assume an EUV light illuminates onto a 40 pairs Mo/Si

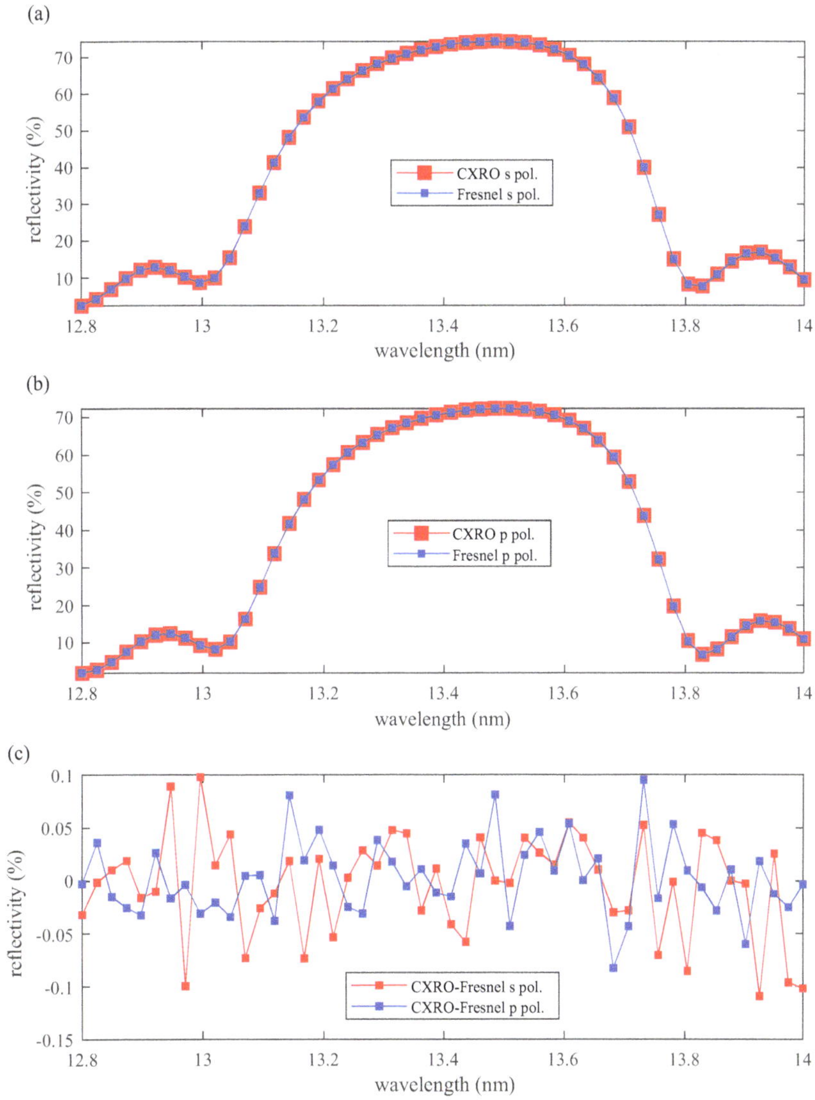

Figure 3.9. Comparisons between the CXRO solver and the in-house Fresnel's equation based solver with 13.5 nm EUV light incident obliquely incidence under (a) *s*-polarization and (b) *p*-polarization. (c) The differences (CXRO—Fresnel) between reflectivity data.

based mirror at 6° with BW = 1.2 nm). The root mean square error (RMSE) for 1D data set as shown below is used to define the data difference of figures 3.9(a) and (b)

$$e_{\mathrm{rms}} = \left[\frac{1}{N_{\mathrm{data}}} \sum_{N_{\mathrm{data}}} (R_{\mathrm{Fresnel}} - R_{\mathrm{CXRO}})^2 \right]^{\frac{1}{2}}, \tag{3.73}$$

where R_{CXRO} is the reflectivity from the CXRO solver, R_{Fresnel} stands for the reflectivity calculated from the Fresnel's equations based solver and N_{data} is the number of total sample points. In figures 3.9(a) and (b), the RMSE between the Fresnel's equations based solver and the CXRO solver for both *s*- and *p*-polarizations are within 4.4%, which is a tolerable difference. The differences (CXRO—Fresnel) between reflectivity data are also plotted in figure 3.9(c). After we verified the reliability of our Fresnel's equations based solver, now we are going to verify the usability of the equivalent layer. By using the Fresnel's equations based solver, we calculate the EUV reflectivity for below structures:

- A 40 Mo/Si ML stack based EUV mirror as shown in figure 3.10(a).
- An equivalent layer based EUV mirror as shown in figure 3.10(b).

There are three calculation cases, in the first case the EUV light illuminates normally onto both of the mirrors, and the results of different pair numbers are shown in figure 3.11(a). In the other two cases the EUV light illuminates obliquely with 6° incident angle onto both of the mirrors, and the results of different pair numbers for *s*- and *p*-polarizations are shown in figures 3.11(b) and (c). The dimension of each layers of the Mo/Si ML stack based EUV mirror can be seen

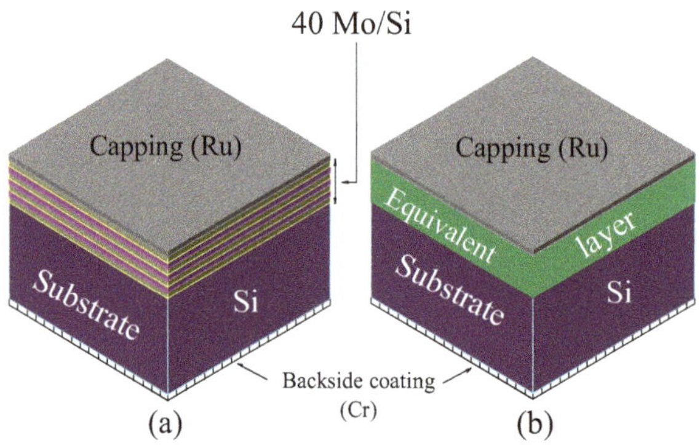

Figure 3.10. (a) A 40 Mo/Si ML stack based EUV mirror and (b) an equivalent layer based EUV mirror.

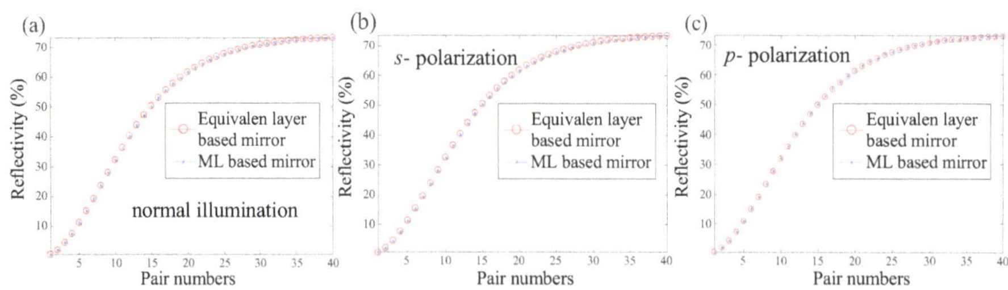

Figure 3.11. Comparisons of calculated reflectivities between the ML based mirror and the equivalent layer based mirror. The results of a 13.5 nm EUV light incident normally are plotted in (a). For oblique illumination at an angle of 6°, the results of *s*- and *p*-polarizations are plotted in (b) and (c).

in table 2.1, and the used refractive indices can be referred to in figure 2.7. As shown in figure 3.11, before stacking up to 20 pairs Mo/Si, the reflectivity increases almost linearly with the increment of the ML pair number. After 20 Mo/Si pairs, the reflectivity growth slows down and eventually reaches a limit. For the normal illumination case, as seen in figure 3.11(a), the reflectivities calculated from the Mo/Si ML based EUV mirror match well with the reflectivities calculated from the equivalent layer based EUV mirror. It shows that the reflectivity of 40 pairs is about 73.28% with equivalent refractive index = 0.0819–*j*0.2357. For an angle of incidence at 6° cases as shown in figures 3.11(b) and (c), the equivalent index of the equivalent layer for *s*- and *p*-polarizations are 0.0797–*j*0.2368 and 0.0892–*j*0.2028. Readers can see that reflectivities calculated from the Mo/Si ML based EUV mirror and from the equivalent layer based EUV mirror at all ML pairs are identical. For *p*-polarization, the reflectivity reaches 73.06% when the Mo/Si ML stacks up to 40 pairs. For *s*-polarization, the reflectivity of 40 pairs Mo/Si is about 74.25%. These data prove that the equivalent layer has identical impedance with the Mo/Si ML, replacing the 40 Mo/Si ML stack with an equivalent layer in FDTD will not have a large impact to numerical accuracy.

3.4.3 Benchmark between rigorous and efficient FDTD methods

To benchmark the efficient FDTD method with the rigorous FDTD method, 2D FDTD calculations of $\lambda = 13.5$ nm EUV light incident normally and obliquely at 6° onto a 40 pairs Mo/Si ML based EUV mirror (rigorous FDTD calculation, refer to figure 3.10(a)) and an one homogeneous equivalent layer based EUV mirror (efficient FDTD calculation, refer to figure 3.10(b)) were performed under different mesh sizes. Both computational time (CPU execution time) and the RMSE of the reflectivity (as compared with numerical result) of efficient FDTD calculations will be benchmarked against the results of rigorous FDTD calculations. The computational domain we used here is 250×600 nm^2 with PBC on the left and the right edges while PML boundaries are on the top and on the bottom edges. Readers can refer to figure 3.12(a) to realize the boundary settings and the dimensions of the computational domain. All calculations were finished on a hexa core 2.60 GHz Intel i7-9750H CPU with 8GB DDR4 RAM. Please note that we extended Si substrate through the PML to simulate infinite thick substrate (there is a huge scale difference between Mo/Si ML and Si substrate, so the substrate is infinite to ML). Four different mesh levels were applied to the calculations, which are 0.675 nm (i.e., $\lambda \times (1/20)$), 0.2085 nm (i.e., Si thickness×1/20), 0.139 nm (i.e., Mo thickness × 1/20) and 0.0695 nm (i.e., Mo thickness × 1/40). The dimension of each of the layers of the 40 pairs Mo/Si ML based EUV mirror are shown in table 2.1, and the refractive indices of Mo, Si, Cr and Ru that are used in FDTD calculations can refer to figure 2.7. For efficient FDTD calculations, the refractive index under EUV normal incidence = 0.0819 – *j*0.2357 (both *s*- and *p*-polarizations). The thickness of the equivalent layer is 278 nm. The refractive indices under angle of incidence 6° are 0.0797 – *j* 0.2368 (*s*-polarizations) and 0.0892 – *j*0.2028 (*p*-polarizations). Readers might notice that the equivalent indices have a very small real part but a larger

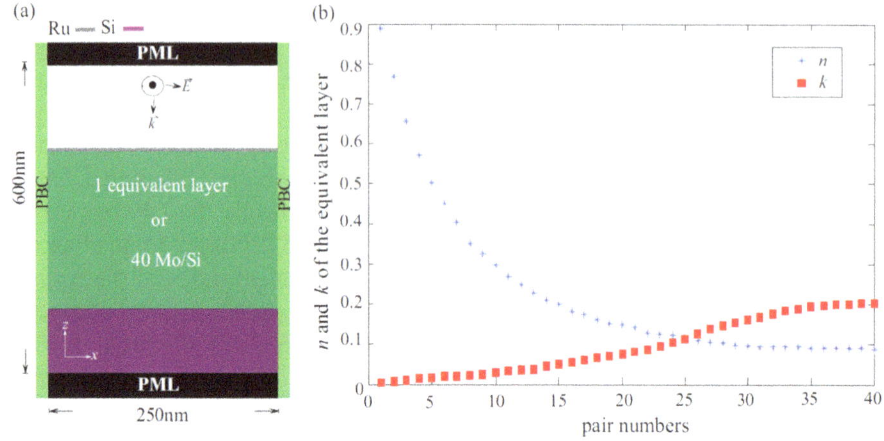

Figure 3.12. (a) Simulation setups of the 2D benchmark tests between rigorous and efficient FDTD calculations. (b) Equivalent refractive indexes of the equivalent Mo/Si for 13.5 nm EUV wavelength. Both *n* and *k* vary with different Mo/Si pair numbers.

imaginary part. The physics behind the index is the phase velocity normally gets faster as adding more film layers, thus, the real part of refractive index gets smaller. In addition, the extinction of the entire ML stack is larger than the one-layer thin film due to the summation of extinction from each film, so it has larger imaginary part for the ML stack. This phenomenon can be observed in figure 3.12(b), which shows that the equivalent refractive indexes vary with adding layer numbers under *s*-polarization. From figure 3.12(b), readers can see the real parts of refractive indices are decreased and the imaginary parts of refractive indices are increased as layer numbers increase. The calculation results are compared in figure 3.13. Note that the analytical solution of reflectivity for a 40 pairs Mo/Si ML based mirror under 13.5 nm EUV light normal incidence with *s*- or *p*-polarization is 73.28%, and the analytical solutions of reflectivities for 6° illumination with *s*- and *p*-polarizations are 74.25% and 73.06%. As seen from figures 3.13(a) and (c), no matter if under normal or 6° illumination, the reflectivities of the efficient FDTD calculations (red dash lines) converges to the theoretical value. The readers can see the reflectivities become more, precisely as the grid size gets finer. In contrast, the reflectivities of rigorous FDTD calculations (blue lines) deviate from the theoretical value with the shrinking of the grid sizes. The root-cause here is because the uniform meshes are not well-aligned with the edges of each film, which is a typical issue of the uniform mesh. To accurately calculate ML problems by using the FDTD method, it requires special meshing treatment to position the grid edges right at the boundary between each films. For instance, to include the level set method [26] in FDTD calculation or to apply a nonuniform grid [8] during the meshing step in the FDTD method. The RMSE reflects the strengths and weaknesses of both cases, as shown in figures 3.13(b) and (d), the analytical results were used as the standard throughout RMS calculations. The RMSEs between reflectivities of rigorous FDTD calculations and analytical results (blue lines) are not converging with the shrinking of the

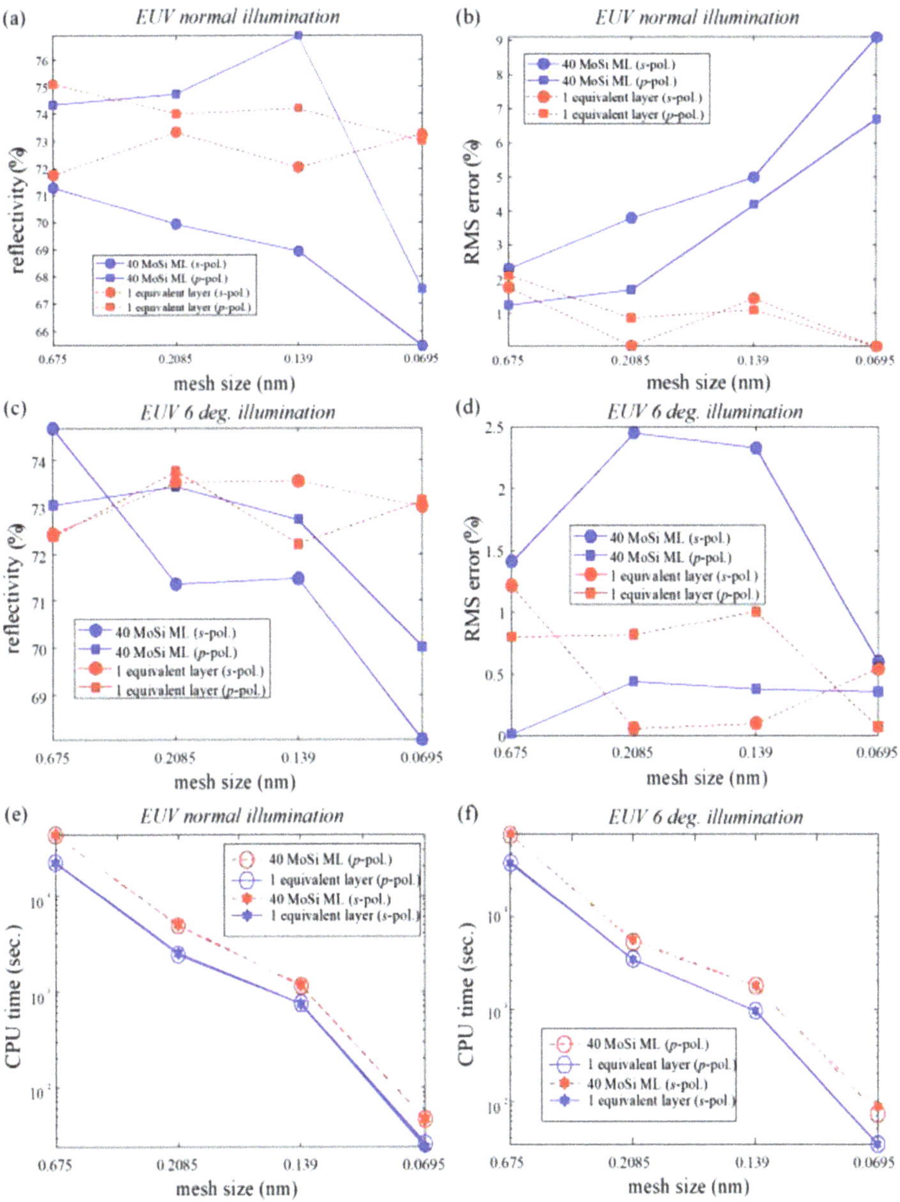

Figure 3.13. Reflectivity and RMSE comparisons of the equivalent layer based EUV mirror and the 40 Mo/Si ML based EUV mirror of FDTD calculations. (a) and (b) are results of normal illumination, while (c) and (d) are results of 6° illumination.

grid size owing to the meshing issue. In contrast, the RMSEs between reflectivities of efficient FDTD calculations and analytical results (red dash lines) trend down as using fine meshes. Besides benchmarking accuracy, we are also interested in calculation efficiency. Figures 3.13(e) and (f) indicate that all computational times are increased as we applied the finer meshes, this applies to all the cases. Moreover, these figures indicate that longer computational time is required to calculate glancing illumination problems rather than normal illumination problems. This is because the optical path length (OPL) of the EUV light in the films becomes longer for glancing illuminations. Thus, the light requires more time to finish the propagation. Mostly, all computational time for efficient FDTD calculations are nearly half of the time as compared with rigorous FDTD calculations. The benchmark results state that the theory can be used to accelerate 2D FDTD calculations for ML based structures with reasonable numerical accuracy. The cases with more complexity will be demonstrated in chapter 4.

References

[1] Erdmann A 2005 Modeling and simulation *Handbook of Photomask Manufacturing Technology* 1st edn ed S Rizvi (San Jose, CA: CRC Press) ch 31

[2] Burger S, Köhle R, Zschiedrich L, Gao W, Schmidt F, März R and Nolscher C 2005 Benchmark of FEM, waveguide and FDTD algorisms for rigorous mask simulation *Proc. SPIE* **5992** 599216

[3] Erdmann A, Kalus C K, Schmöller T and Wolter A 2003 Efficient simulation of light diffraction from three-dimensional EUV masks using field decomposition techniques *Proc. SPIE* **5037** 482–93

[4] Domnenko V, Klimpel T, Viehoever G, Koop H, Melvin III L S and Schmoeller T 2007 OPC verification on cell level using fully rigorous mask topography simulation *Proc. SPIE* **6730** 67304J

[5] Clifford C H and Neureuther A R 2007 Fast three-dimensional simulation of buried EUV mask defect interaction with absorber features *Proc. SPIE* **6730** 67301s

[6] Clifford C H and Neureuther A R 2009 Fast simulation methods and modeling for extreme ultraviolet masks with buried defects *J. Micro/Nanolith. MEMS MOEMS* **8** 0314021

[7] Yee K 1966 Numerical solution of initial boundary value problems involving Maxwell's equations in isotropic media *IEEE Trans. Antennas Propag.* **14** 302–7

[8] Taflove A and Hagness S C 2005 Introduction to Maxwell's equation and the Yee algorithm *Computational Electrodynamics: The Finite-Difference Time-Domain Method* 3rd edn ed T Millogan (Boston, MA: Artech House) ch 3

[9] Kelvin–Stokes theorem from Wikipedia (https://en.wikipedia.org/wiki/Kelvin%E2%80%93Stokes_theorem)

[10] Nyquist H 1932 Regeneration theory *Bell System Tech. J.* **11** 126–47

[11] Courant R, Friedrichs K and Lewy H 1928 Über die partiellen Differenzengleichungen der mathematischen Physik *Math. Ann.* **100** 32–74

[12] Zheng G, Kishk A A, Glisson A W and Yakovlev A B 2006 Implementation of Mur's absorbing boundaries with periodic structures to speed up the design process using finite-difference time-domain method *Prog. Electromagn. Res.* **58** 101–14

[13] Sudiarta I W 2003 An absorbing boundary condition for FDTD truncation using multiple absorbing surfaces *IEEE Trans. Antennas Propag.* **51** 3268–75

[14] Bérenger J P 2015 A historical review of the absorbing boundary conditions for electromagnetics *Forum Electromagn. Res. Methods Appl. Technol.* **9** 1–28

[15] Moerloose J D and Stuchly M A 1996 An efficient way to compare ABCs *IEEE Antennas Propag. Mag.* **38** 71–5

[16] Rappaport C.M and Winton S 2000 Using the PML ABC for air/soil wave interaction modeling in the time and frequency domains *Subsurf. Sens. Technol. Appl.* **1** 289–303
[17] Bérenger J P 1994 A perfect matched layer for the absorption *J. Comput. Phys.* **114** 185–200
[18] Sacks Z S, Kingsland D M, Lee R and Lee J F 1995 Performance of an anisotropic artificial absorber for truncating finite element meshes *IEEE Trans. Antennas Propag.* **43** 1460–3
[19] Entegris (https://www.entegris.com/shop/en/USD/Products/Mask-and-Reticle-Handling/Extreme-Ultraviolet-Light-%28EUV%29-Reticle-Pods/c/extremeultravioletlighteuvreticlepods)
[20] Yamaguchi T, Ikuta H, Tomofuji T, Yanagi Y, Itoh Y, Oka T and Mizutani U 2008 Reflective properties of Mo–Si multilayer for EUV lithography deposited by the magnetron sputtering device with superconducting bulk magnets *Phys.* C **468** 2170–3
[21] Hiruma K, Miyagaki S, Yamanashi H, Tanaka Y and Nishiyama I 2008 Performance and quality analysis of Mo-Si multilayers formed by ion beam and magnetron sputtering for extreme ultraviolet lithography *Thin Solid Films* **516** 2050–7
[22] Ono Y 2006 Transmittance analysis of three-dimensional photonic crystals by the effective medium theory *Appl. Opt.* **45** 131–6
[23] Torrungrueng D and Lamultree S 2008 Analysis of planar multilayer structures at oblique incidence using an equivalent BCITL model *Progr. Electromagn. Res.* C **4** 13–24
[24] Chen D K 1989 Theory and applications of transmission lines *Field and Wave Electromagnetics* 2nd edn (Boston, MA: Addison-Wesley) ch 9
[25] The Center for X-Ray Optics (CXRO) (http://cxro.lbl.gov/) (accessed 19 September 2020)
[26] Olsson E and Kreiss G 2005 A conservative level set method for two phase flow *J. Comput. Phys.* **210** 225–46

Chapter 4

Simulation cases

In this chapter, the near-field and far-field results of different kinds of extreme ultraviolet (EUV) mirrors will be calculated by using the finite-difference time-domain (FDTD) method. There are two critical key performance indicators to EUV mirror design, one is mirror reflectivity and the other is near-field scattering pattern. Although the reflectivity is more relative to the throughput of the EUV scanner, the near-field scattering pattern should also be considered since it concerns image quality. From the near-field scattering patterns, designers foresee how patterns are formed on the wafer. For all the cases in sections 4.1–4.3, we will perform calculations by using both the efficient FDTD and the rigorous FDTD approaches for benchmark comparison.

4.1 2D simulation: EUV mirror with surface roughness

A mirror containing surface roughness/unflatness has attracted our attention, this is the first case we would like to demonstrate here. Normally, this realistic situation results from out-of-plane distortion (OPD) of the mirror substrate, which has been discussed in section 2.1.1. On that occasion, the mirror suffers unbalanced Mo/Si layers deposition so that the roughness/unflatness will be formed on top of the mirror surface. How the roughness/unflatness above the surface influences the reflectivity of the EUV mirror owing to additional light scattering was numerically studied in [1], this is one critical technical item that needs to be considered in a EUV exposure system. What we are also interested in is if the high efficiency calculation results of the efficient FDTD method is still valid for the surface roughness/flatness mirror. In this subsection we are going to calculate that the mirror contains surface roughness in 2D by using the rigorous FDTD and efficient FDTD methods. As seen in figures 4.1(a) and (b), we designed 1D periodic Si gratings above the mirror surface for both the 40 pairs Mo/Si ML based EUV mirror (rigorous FDTD case) and the one-layer equivalent layer based EUV mirror (efficient FDTD case) to approximate the surface roughness situation. The 1D Si gratings are assumed to have uniform

doi:10.1088/978-0-7503-2652-0ch4

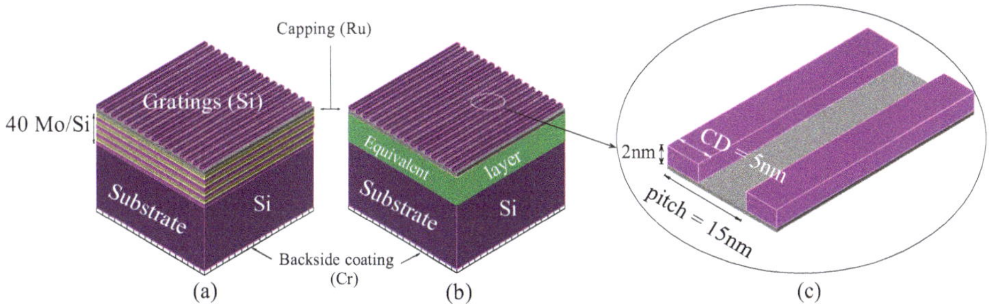

Figure 4.1. 1D periodic Si gratings on top of (a) a Mo/Si ML based EUV mirror and (b) an equivalent layer based EUV mirror. (c) A zoom-in of the periodic 1D gratings.

thickness, linewidth and pitch. In this case it has 5 nm linewidth (CD = 5 nm), 15 nm pitch and 2 nm thickness, as shown in figure 4.1(c). The computational layout to calculate these two cases can be seen in figure 4.2(a). The total computation domain is 250×600 nm^2 with a 0.2085 nm grid size resolution. Using the same theory as section 3.4.3, we extend the Si substrate through the perfect matched layer (PML) to simulate infinite thick substrate. Besides, periodic boundary conditions (PBCs) were applied at both the left and right sides as we assume gratings repeat along the x-direction. As for the light source, we assumed there was a pure EUV light with a central wavelength (CW) = 13.5 nm illuminating the mirrors obliquely at an angle of incidence 6°. The refractive indexes and extinction coefficients of Si, Mo, Ru capping and the backside coating Cr at λ = 13.5 nm that were used in our calculations can be seen in figure 2.7. All calculations were finished on a hexa core 2.60 GHz Intel i7–9750H CPU with 8 GB DDR4 RAM.

The electric field distributions of both rigorous and efficient FDTD cases are provided in figures 4.2(b) and (c) (p-polarization). In both figures, the clear EUV interference fringe above the mirror surface can be easily observed, this is formed due to incoming waves superposed with reflection waves. The interference fringes above the mirror surface in figures 4.2(b) and (c) are on the same intensity scale, this demonstrates that the equivalent layer is able to reflect the same amount of EUV light just like the Mo/Si ML. Nevertheless, the EUV interference fringe inside the Mo/Si ML can only be seen in figure 4.2(e) owing to light interferences of the Mo and Si ML. In contrast, no interference inside the equivalent layer can be observed as shown in figure 4.2(f). The interference fringe of the electric field intensity inside the ML contributes major differences between the Mo/Si ML based mirror and the equivalent layer based mirror. This interference fringe resulted from multiple EUV reflections and transmissions within the Mo/Si pairs, meaning the FDTD calculations of EUV propagation within the Mo/Si ML (rigorous FDTD method) are more challenging than the FDTD calculations within the equivalent layer (efficient FDTD method) because multiple reflections and transmissions induced by ML need to be calculated. This is where the rigorous FDTD method loses the numerical accuracy and produces longer computational time. Figure 4.2(d) demonstrates the squared error of the electric intensity between rigorous and efficient FDTD calculations, only weak intensity differences below the mirror surface can be

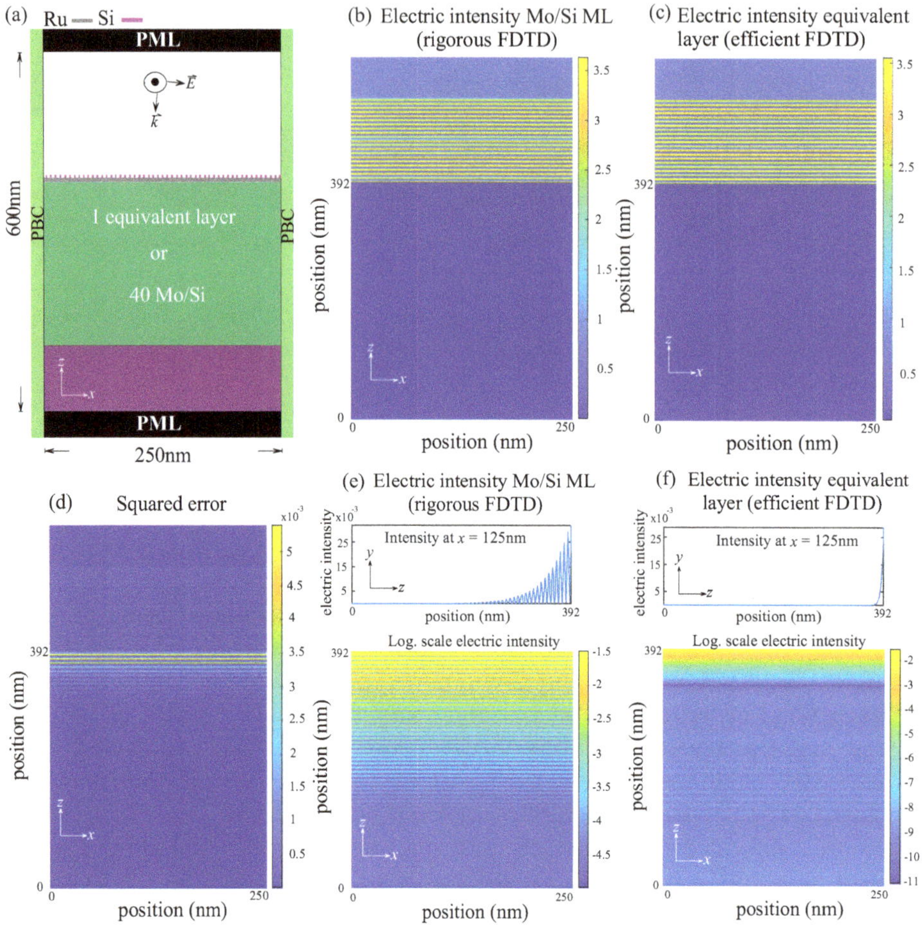

Figure 4.2. Demonstrations of 2D rigorous and efficient FDTD calculations for the periodic Si gratings added EUV mirror under 13.5 nm EUV light incident obliquely at an angle of 6°. (a) Computational layout. (b) and (c) Calculated electric intensity from rigorous FDTD and efficient FDTD methods (*p*-polarization). (d) Squared error between figures (b) and (c). (e) and (f) The results of z = 0–392 nm region in (b) and (c), in which the upper plots stand for the results at x = 125 nm, the lower plots are log scale intensities.

observed. However, these slight differences won't affect the design since the intensity scale of the maximum squared error between efficient FDTD calculation and rigorous FDTD calculation is below 0.6%. Based on figure 4.2(d), we can also justify that the reflectivity calculations from rigorous and efficient FDTD methods are very close since the maximum squared error above the mirror is below 0.1%.

Mostly, the computational time comparisons between the rigorous FDTD calculation and the efficient FDTD calculation for the above cases are show in table 4.1. Just like the conclusion we made in section 3.4.3, the computational time of the efficient FDTD calculations are half of the computational time of the rigorous FDTD calculations. Concerning numerical accuracy of the efficient FDTD calculation, if we take the electric field intensity from the rigorous FDTD calculation as

Table 4.1. Efficiency comparisons between the efficient FDTD calculations and the rigorous FDTD calculations for the 40 Mo/Si ML based EUV mirror covered by 1D periodic Si gratings under EUV incident obliquely at angle of 6°.

Methods	Rigorous FDTD		Efficient FDTD	
Polarization	*s*-	*p*-	*s*-	*p*-
CPU time (s)	1141	1160	509	512

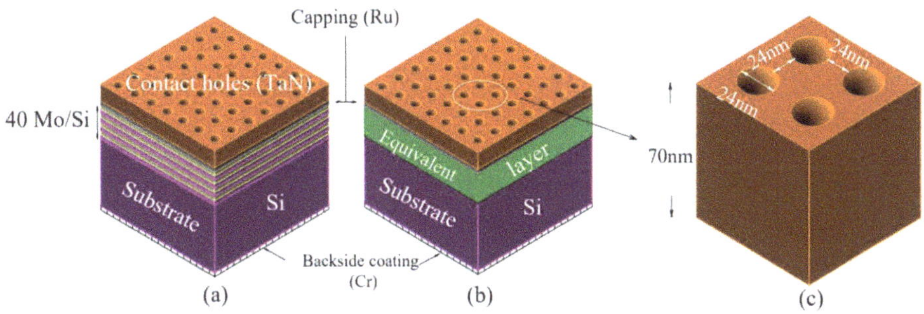

Figure 4.3. Periodic contact holes on top of (a) a Mo/Si ML based EUV mirror and (b) an equivalent layer based EUV mirror. (c) A zoom-in of the periodic contact holes.

the reference, the root mean square error (RMSE) between rigorous FDTD calculations and efficient FDTD calculations for both *s*- and *p*-polarizations are 0.24 and 0.26%, respectively. We conclude that to simulate an EUV mirror with periodic surface roughness in 2D, the efficient FDTD method works better than the rigorous FDTD method in consideration of CPU time.

4.2 3D simulation: EUV mirror patterned with periodic contact holes

An important feature in the integrated circuit (IC) manufacturing process, mentioned in section 1.3, which are the contact holes, is that they must be created within the phosphosilicate glass (PSG) and the undoped silicate glass (USG) layers during the back-end-of-line (BEOL) process. The contact holes are openings on the insulation film that are used to connect the layers and the interconnect layers. The contact holes must be firstly patterned on the EUV mirror (so it becomes reticle) then the pattern must be transferred from the EUV mirror to the wafer by means of EUV lithography. In this section, we are going to demonstrate how the FDTD calculation is used to predict the performance of the contact holes patterned EUV mirror.

We define periodic contact holes on a TaN layer coated EUV mirror, calculated with rigorous 3D FDTD method and efficient 3D FDTD method, as shown in figures 4.3(a) and (b). The distance between each contact hole center are designed to be equal. In this case each hole has 24 nm diameter width, 24 nm edge-to-edge distance and 70 nm depth, as shown in figure 4.3(c). For the EUV light source setting in the FDTD calculations, according to [2], the bandwidth (BW) of the EUV light source is 0.8 nm, which is roughly ranging from 13.1 nm to 13.9 nm if the CW = 13.5 nm is applied. The EUV source intensities verses wavelength can be estimated by a Gaussian function as

$$g(\mathrm{x}) = Ae^{\frac{-(\lambda-13.5)^2}{2F^2}}, \tag{4.1}$$

where A stands for intensity at the center of the EUV light source, λ is the wavelength, and the parameter F is related to the full width at half maximum (FWHM), which is the general BW definition of a Gaussian pulse, which is determined as

$$\text{FWHM} = 2(2\ln 2)^{\frac{1}{2}} F. \tag{4.2}$$

The illumination optics of an EUV scanner contains typically four mirrors, from the laser produced plasma light source (LPPS) to the reticle as seen in figure 1.8, small tilted angles (typically 6°) are applied to each mirror [3]. For this reason the light intensity and the BW can be diminished significantly when arriving at the reticle. To fulfill this requirement in the 3D FDTD calculation, we defined the CW and the BW of the source emitted from the LPPS to be 13.5 nm and 0.8 nm respectively. The profile is depicted in the red line in figure 4.4. By applying equation (4.2), the profiles of the BW and intensity diminished EUV source after experiencing M_1–M_4 are all depicted in figure 4.4(a). The black line indicates the proper power and BW of the EUV source to illuminate the reticle. To show readers how the structures were constructed in the computational domain, the computational layouts of the x–z-plane (y = 24 nm) and the x–y-plane (z = 462 nm) for both rigorous and efficient FDTD calculations are shown in figures 4.4(b) and (c). From these two figures readers should be able to observe the total computation domain is 48 × 48 × 600 nm^3. We performed the same calculations with two different grid size resolutions, which are $\Delta x = \Delta y = \Delta z = 0.675$ nm and 0.2085 nm. The left and the right boundaries in figure 4.4(b) are defined to be the PBC to simulate repeated EM scatterings along the x-direction owing to periodic contact holes. We extended Si substrate through the bottom PML to simulate infinite thick substrate. The same reason in figure 4.4(b), PBCs were applied to encompass the whole simulation domain in figure 4.4(c) (figures 4.3 and 4.4).

Based on the black line in figure 4.4(a), we applied a Gaussian pulse with CW = 13.5 nm and BW = 0.3 nm as our source in our FDTD calculations (λ = 13.35–13.65 nm). Table 4.2 lists the equivalent refractive index of the equivalent layer as indicated in figure 4.3(b) corresponding to wavelength ranging from 13.35 to 13.65 nm for s- and p-polarizations used in the calculations. Before using the FDTD method to calculate the near-field scattering results, what designers normally do is to quickly estimate the reflectivity beforehand. This is to ensure the designed mirror fulfills the required reflectivity before we spend effort for FDTD calculations since this is one of the critical key performance indicators (KPIs). This can quickly be done by using the Fresnel's equations based solver described in section 3.4.2. By using this solver together with the indexes provided in table 4.2 and figures 2.7 and 4.5, we are able to do a quick reflectivity assessment for the periodic contact holes covered EUV mirror, and this estimation will not take too much effort. In our contact holes design,

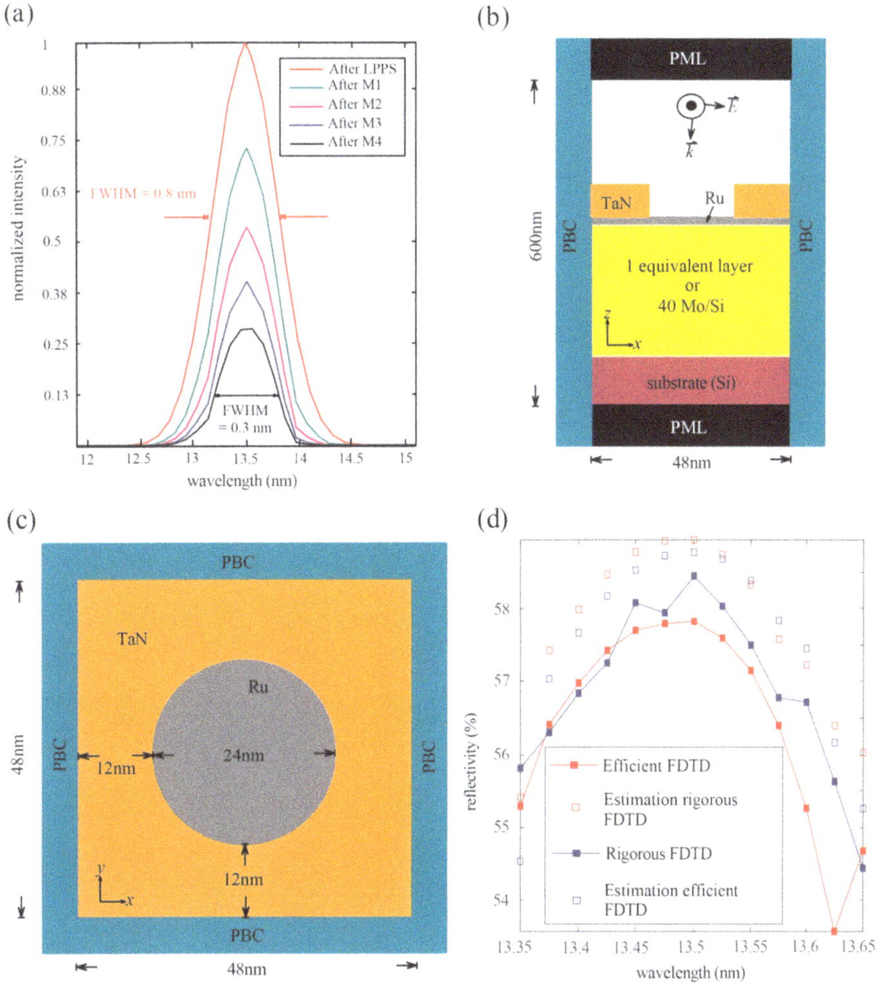

Figure 4.4. (a) Gaussian-shaped EUV source intensities versus wavelengths after different times reflection, in which the EUV intensity after experiencing M_4 mirror as shown in the black line are applied to 3D FDTD calculations for periodic contact holes covered EUV mirror. (b) and (c) Simulation setups of 3D FDTD calculations for periodic contact holes covered EUV mirror in x–z and x–y views. (d) Reflectivity of periodic contact holes covered EUV mirror, in which the hollow red squares and the blue squares are the reflectivities of the ML and the equivalent layer based mirrors from Fresnel estimation, while the red and the blue solid squares are the reflectivities of the ML and the equivalent layer based mirrors from 3D FDTD calculation.

the holes–TaN mix ratio is 19.6%. This ratio is used to calculate the equivalent indexes of contact holes contained TaN (TaN_{eq} as shown in figure 4.5). Here are a few remarks we would like to make for figure 4.5. In figure 4.5(a), the refractive index of the TaN_{eq} is always closer to 1 compared to the index of the TaN over all the wavelengths. This totally makes sense because TaN_{eq} in principle is treated as a new substance containing 19.6% free space, thus the light speed in the holes–mixed TaN should be closer to the light speed in free space, which results in fewer light ray bends

Table 4.2. Equivalent refractive indexes of a 40 pairs Mo/Si ML stack for wavelength ranging from 13.35 to 13.65 nm.

Wavelength (nm)	13.35	13.375	13.4	13.425	13.45	13.475	13.5	13.525	13.55	13.575	13.6	13.625	13.65
Equivalent index (*s*-pol.)	0.0883-j0.2012	0.0878-j0.2194	0.085-j0.2191	0.0841-j0.2271	0.081-j0.2471	0.0819-j0.2372	0.079-j0.2368	0.0815-j0.2372	0.083-j0.2326	0.0868-j0.2367	0.087-j0.2345	0.0923-j0.2432	0.094-j0.2498
Equivalent index(*p*-pol.)	0.0983-j0.2684	0.0969-j0.2735	0.0939-j0.2768	0.0919-j0.2051	0.0905-j0.1999	0.0896-j0.1979	0.0892-j0.2028	0.0898-j0.1991	0.0911-j0.3135	0.0935-j0.2078	0.0955-j0.3061	0.1015-j0.2143	0.1103-j0.2089

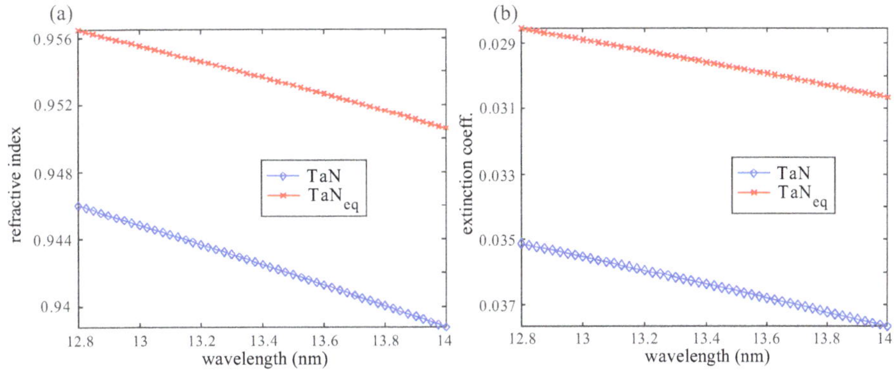

Figure 4.5. Refractive indexes and extinction coefficients of TaN and equivalent TaN (TaN_{eq}, contains 19.6% contact holes).

towards normal. Besides, the extinction coefficient of the TaN_{eq} is always closer to 0 compared to the extinction coefficient of the TaN over all the wavelengths owing to the same physics as shown in figure 4.5(b). The estimated reflectivities of both rigorous FDTD and efficient FDTD calculations are plotted as hollow red squares and hollow blue squares in figure 4.4(d). In this figure only *s*-polarization is estimated since there will not be too many differences between *s*- and *p*-polarization from the reflectivity aspect. Based on the estimations in figure 4.4(d), the peak reflectivity for both estimations appears at 13.5 nm (~59%), which explains the reason for CW selection.

As mentioned before, we performed 3D FDTD calculations under a grid size = 0.675 nm and grid size = 0.2085 nm in order to benchmark the scattering results and the CPU time between the efficient and the rigorous FDTD methods. For such fine meshes, it would be impossible to complete 3D calculations on a single CPU due to both memory and computational time limitations. Thus, these cases were calculated with multiple CPUs through a message passing interface (MPI) [4]. The calculated electric scatterings on top of the TaN absorber (*x*–*y*-plane) are provided in figure 4.6. Please note that the observation plane for all calculations here is right above the TaN absorber, the electric field distribution above the TaN absorber provides the information concerning the resist pattern. The results for *s*-polarization are provided in figures 4.6(a)–(j) and the results for *p*-polarization are provided in figures 4.6(k)–(t). For *s*-polarization results, figures 4.6(a) and (c) are the intensities from rigorous FDTD calculations with 0.675 nm mesh size for monochromatic 13.5 nm and 13.375 nm light illuminations, while results for 0.2085 nm mesh size are provided in figures 4.6(b) and (d). Correspondingly, figures 4.6(f) and (h) are the intensities from efficient FDTD calculations with 0.675 nm mesh size for monochromatic 13.5 nm and 13.375 nm light illuminations, while results for 0.2085 nm mesh size are provided in figures 4.6(g) and (i). The *s*-polarized results of monochromatic illumination indicate that the electric intensity is not evenly distributed but is mostly intensified above and below the contact hole. In addition, this phenomenon becomes more recognizable when applying finer mesh as seen in

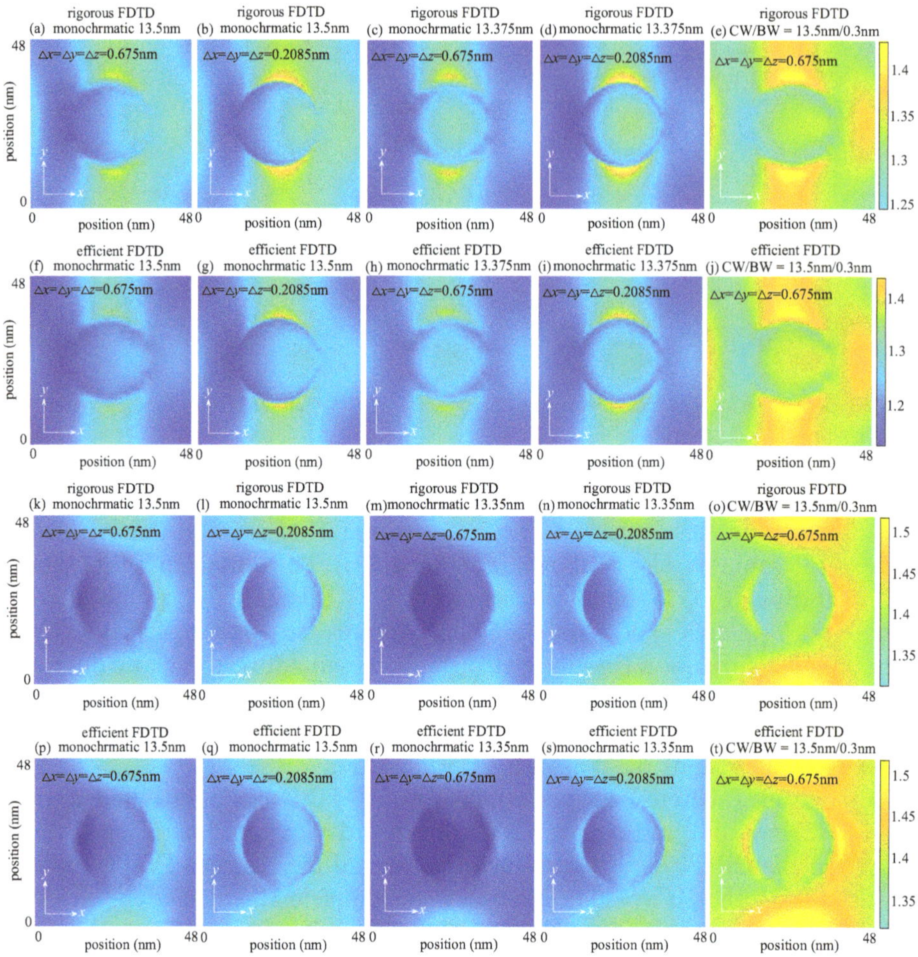

Figure 4.6. Calculated electric field intensities right above the TaN absorber from 3D rigorous FDTD calculations and efficient FDTD calculations, in which (a)–(j) are *s*-polarized results and (k)–(t) are *p*-polarized results.

figures 4.6(b), (d), (g) and (i). It seems that in this case the fine mesh provides more accurate results because the contour of the results from 0.2085 nm meshes corresponds with the results from 0.675 nm meshes. Specifically, the results from 0.2085 nm meshes provide better resolution. Considering the tradeoffs by applying fine mesh, the results from 0.675 nm offer sufficient information to anticipate how the periodic contact holes pattern will be transferred onto the wafer. Similar to results from monochromatic illuminations, the *s*-polarized results from wideband illumination (CW/BW = 13.5 nm/0.3 nm) for both rigorous and efficient FDTD calculations appear with the same trend as monochromatic results and can refer to figures 4.6(e) and (j). All the results from wideband illumination indicate that most of the electric intensities are gathered at the regions above and below the contact hole. Besides, both the right and the left sides close to the border of the computational domain also get obviously intensified in wideband results.

To recap briefly for the results from wideband illumination, firstly, the overall field intensity behaves stronger than the results from any monochromatic illuminations since they are in theory a superposition of monochromatic results. Secondly, all the characteristics appearing in the results of monochromatic illumination can also be seen in wideband results. From figures 4.6(a)–(j), it doesn't matter which wavelength is used, the results between rigorous and efficient FDTD match well. For *p*-polarization results as shown in figures 4.6(k)–(t), λ = 13.5 nm and 13.35 nm are selected as our monochromatic illumination examples. Figures 4.6(k) and (m) are the intensities from rigorous FDTD calculations with 0.675 nm mesh size for monochromatic 13.5 nm and 13.35 nm light illuminations, while results for 0.2085 nm mesh size are provided in figures 4.6(l) and (n). Correspondingly, figures 4.6(p) and (r) are the intensities from efficient FDTD calculations with 0.675 nm mesh size for monochromatic 13.5 nm and 13.35 nm light illuminations, while results for 0.2085 nm mesh size are provided in figures 4.6(q) and (s). The *p*-polarized results have different contours compared to the *s*-polarized results, it doesn't matter if you observe the contour from monochromatic or from wideband illuminations. As seen from figures 4.6(k), (m), (p) and (r), the intensity is roughly rotated by 90° compared with figures 4.6(a), (c), (f) and (h) per Brewster's law [5]. Even so, they are not exactly the outcomes from only having 90° rotation owing to 6° tilted illumination. The intensity is intensified at the left and the right edges of the contact hole, and the areas are relatively small compared to the intensified areas above and below the contact hole of figures 4.6(a), (c), (f) and (h). Same to *s*-polarization, this phenomenon becomes more distinguishable when applying finer mesh as seen in figures 4.6(l), (n), (q) and (s) because 0.2085 nm meshes provide better resolution than 0.675 nm meshes. For wideband illumination results as shown in figures 4.6(o) and (t), the results (for both rigorous and efficient FDTD calculations) specify that most of the electric intensities are gathered at the left and the right edges of the contact hole. This shows the same trend as monochromatic results, which is the same effect as *s*-polarized results. Besides, both the top and the bottom areas close to the border of the computational domain also are obvious intensified in both wideband results, and they are formed per Brewster's law by nature.

To recap the *p*-polarized wideband results, the overall field intensity behaves stronger than the results from any monochromatic illuminations as they are also a superposition of monochromatic results. In addition, all the characteristics of monochromatic illumination results are captured in wideband results. Mostly, the results between rigorous and efficient FDTD match well, as illustrated in figures 4.6(k)–(t). The calculated reflectivities from both rigorous and efficient FDTD under different monochromatic illuminations are plotted as solid squares in figure 4.4(d), in which the red solid squares stand for reflectivities from efficient FDTD calculations and the blue solid squares stand for reflectivities from rigorous FDTD calculations. The reflectivities estimated from Fresnel's equations based solver are always higher than the reflectivities from FDTD calculations although the differences are not huge. This is because the Fresnel's equations based solver utilize a

Table 4.3. Computational efficiency comparisons between 3D efficient FDTD calculations and rigorous FDTD calculations for contact holes patterned mirror under EUV incident obliquely at angle of incidence of 6°.

Polarization			*s*-		*p*-	
Mesh size (nm)			0.675	0.2085	0.675	0.2085
CPU time (s)	Rigorous FDTD	$\lambda = 13.5$ nm	1758	58 014	1832	57 833
		$\lambda = 13.375$ nm	1885	61 506	1892	60 892
	Efficient FDTD	$\lambda = 13.5$ nm	985	28 373	974	27 993
		$\lambda = 13.55$ nm	954	29 884	988	29 044

more ideal model than FDTD calculations, the solver assumes flat layers are calculated thus ignores the geometry of the contact holes. Hence, the estimated reflectivities between wavelengths are relatively smooth compared to the reflectivites from FDTD calculations. The computational efficiency comparisons between 3D rigorous FDTD and efficient FDTD calculations for contact holes patterned mirror are provided in table 4.3. From the table, what we see is no matter which FDTD method was applied, the CPU time of the rough mesh (0.675 nm) is always shorter than the CPU time of the fine mesh (0.2085 nm). Specifically, it takes 32ish times longer to complete the calculation using the fine mesh for all cases. As for efficiency comparisons between rigorous and efficient FDTD methods, no matter which meshes were applied, efficient FDTD calculations are almost two times faster than rigorous FDTD calculations. To sum up, we conclude that the efficient FDTD calculation can be a useful tool to accelerate the full 3D calculation for periodic contact holes patterned mirror with reasonable numerical accuracy. It not only reduces the CPU time but also has accuracy with tolerable numerical errors.

4.3 3D simulation: porous EUV mirror

During the manufacturing processes of the ML stack of the EUV mirror, Mo and Si thin film layers undergo the layer-adding process, which causes surfaces unflatness (or rough surfaces) between the layers. The Mo/Si ML stack can be fabricated using a radio frequency/direct current magnetron sputtering system with desired substrate temperature and working power. Working pressure, substrate temperature, and working power should be adjusted to fulfill the deposition requirements. Afterward, the sputtered Mo (or Si) atoms are transported to the substrate through a region of reduced pressure and are condensed on the substrate. After each ion implantation, the surface unflatness and cracks in layers are therefore formed due to thermal effects. The surfaces unflatness results in either additional intermixing layers that are mixed with both Mo and Si atoms or porous Mo/Si layers. In the past the porous Mo/Si ML was treated as defects from imperfect manufacturing and people tried to minimize these effects in order to achieve smooth surfaces and uniform layers [6, 7]. Unintentionally, some researchers realized that these imperfections in the EUV mirror could be beneficial to the throughput of EUV lithography because it is possible to enhance the reflectivity by eliminating the loss tangent by nature that exists in materials. Thus, this resulted in other research fields to investigate how the desired porous Mo/Si stack can be fabricated by applying current technology [8–10]. It was found that by adding the porous Si layers into the Mo/Si ML stack, the reflectivity of the EUV mirror can be extraordinary enhanced [11]. The scientific background of using porous Si is that as EUV radiation interacts with a Mo/Si ML stack, the superposition of the incident and reflected EUV generates a standing wave field in the ML stack. As the Mo has a greater extinction coefficient than the Si in the 13.5 nm wavelength region (refer to figure 2.7), the Mo is designed as the absorber to position into the low intensity standing wave field. Then, other high intensity regions will be filled with Si layers.

In this chapter, we take the EUV mirror with porous Si layers as an example to demonstrate how the scattering effects and the reflectivity can be calculated through the rigorous 3D FDTD method and efficient 3D FDTD method, as indicated in figures 4.7(a) and (b) respectively. For demonstration purposes, we are not going to consider complex porous in our calculations. Instead, we approximate porous Si layers by considering spherical bubbles with a uniform arrangement within each Si layer of the 40 Mo/Si ML stack, as seen in figure 4.7(c). The size and distance between each bubble center are designed to be equal. All the bubbles in each Si layer have equal position in the *z*-direction. In our cases the diameter of the bubbles in the Si layers are assumed to be 2.08 nm, 2.5 nm, 2.9 nm, 3.3 nm and 3.7 nm, meaning the corresponding ratios of bubble to Si are 6.5%, 11.3%, 17%, 26.8% and 38.1%. In our calculations, the light source is a monochromatic 13.5 nm EUV will 6° illumination. The calculation layouts are illustrated in figure 4.8. The total computational domain is $4.17 \times 4.17 \times 600\ \text{nm}^3$. For rigorous FDTD calculation, the dimensions of each layer can be refer to in table 2.1. We performed calculations under grid size = 0.675 nm. The same as section 4.3, it would be tough to complete 3D calculations for such fine meshes on a single CPU. Thus, all cases in this chapter were calculated with multiple CPUs through MPI. The left and the right boundaries in figure 4.8(a) are defined to be PBC to simulate repeated EM scatterings along both the *x*- and *y*-directions owing to the periodic arrangement of the bubbles in Si. Thus, only one bubble in each layer is sufficient. We extend the Si substrate through the bottom PML to simulate infinite

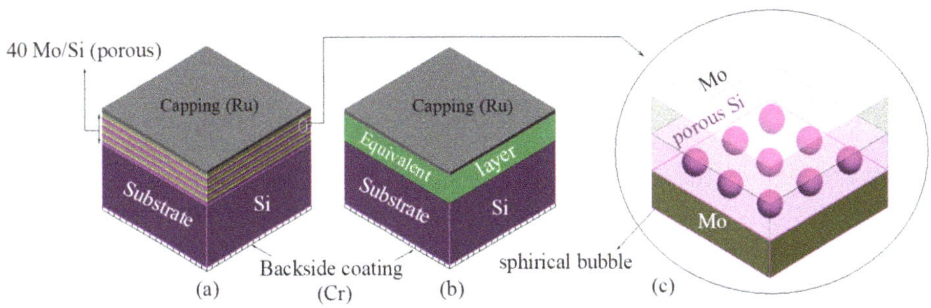

Figure 4.7. Illustrations of FDTD calculations for (a) 40 Mo /Si (porous) ML based EUV mirror and (b) an equivalent layer based EUV mirror. (c) A zoom-in of a porous Si layer approximated by defining spherical bubbles with periodic arrangement.

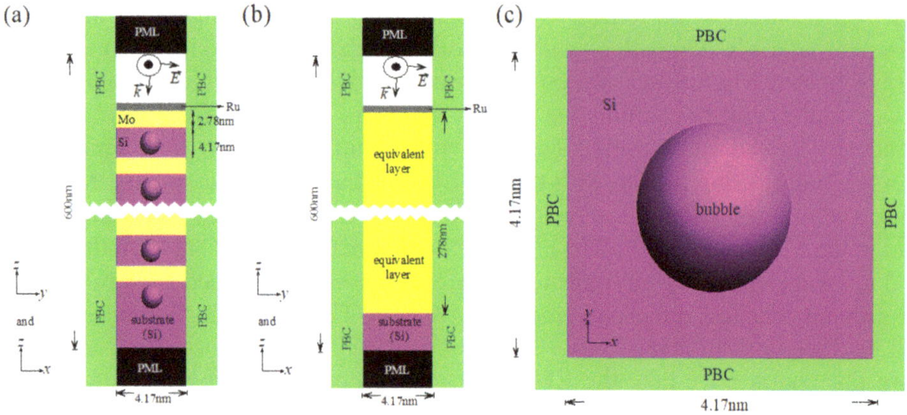

Figure 4.8. Simulation setups of 3D FDTD calculations for (a) 40 Mo /Si (porous) ML and (b) one equivalent layer based EUV mirrors in *x*–*z* and *y*–*z* views. (c) Simulation setups of 3D FDTD calculations for 40 Mo /Si (porous) ML based EUV mirror in the *x*–*y* view, this is the plane at the center of every Si layer in the *z*-direction.

thick substrate. For every bubble ratio, the center of the bubble is overlapped with the center of mass of the Si layer, and the bubbles only exist in Si layers. The configuration of the computational domain the same as the rigorous FDTD calculation is applied to the efficient FDTD calculation, as shown in figure 4.8(b). They share the same boundaries, thicknesses of Ru and substrate, size of the computational domain and light source. The only difference is the setup of the Mo/Si ML stack. Instead of putting 40 Mo and 40 Si films in the computational domain, the efficient FDTD calculations use one equivalent layer combining 40 Mo/Si_{eq} (Si contains bubble) with 278 nm thick. The concept of converting a 40 Mo/Si (porous) ML stack into an equivalent layer is illustrated in figure 4.9(a). At the beginning each porous Si layer is converted into an equivalent Si layer (Si_{eq}), so that the 40 pairs Mo/Si (porous) ML stack becomes a 40 pairs Mo/Si_{eq} ML stack. Afterward, by applying the equivalent layer approach as introduced in section 3.4 to the 40 pairs Mo/Si_{eq} ML stack, the equivalent indices of a one equivalent layer derived from a 40 pairs Mo/Si_{eq} with 278 nm thick are established. The equivalent indices for different bubble ratio under 13.5 nm EUV are plotted in figure 4.9(b) and (c). From these two plots, it can be seen that as the bubble ratio increases, the extinction coefficient of both *s*- and *p*-polarizations decreases. These values have a tendency towards high EUV reflectivity. For equivalent refractive indices, the value moves toward 1 as the bubble ratio is increased. This makes sense because the refractive index of the bubble is 1. Besides, this means the waves can be propagated much more faster than in vacuum. Thus, this accelerates the efficient FDTD calculation. Per the setups in figure 4.8, the calculated electric field intensities (*s*-polarized) above the mirror (*x*–*y*-plane) under different bubble ratios from both rigorous FDTD and efficient FDTD calculations are provided in figure 4.10, in which figures 4.10(a)–(e) are results of rigorous FDTD calculations and figures 4.10(f)–(j) are results of efficient FDTD calculations. It can be seen from

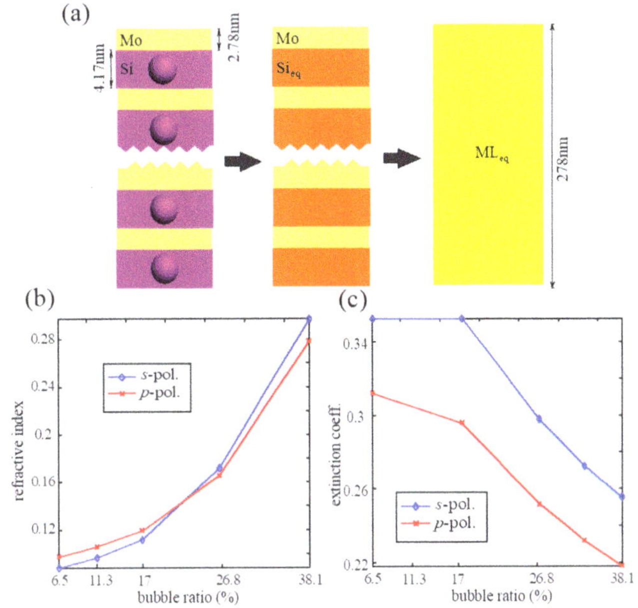

Figure 4.9. (a) Convert a 40 pairs Mo/Si (porous) ML stack into an equivalent layer to be used in figure 4.8(b). (b) and (c) Equivalent refractive indexes and extinction coefficients of the equivalent 40 pairs Mo/Si (porous) ML stack for different bubble ratios.

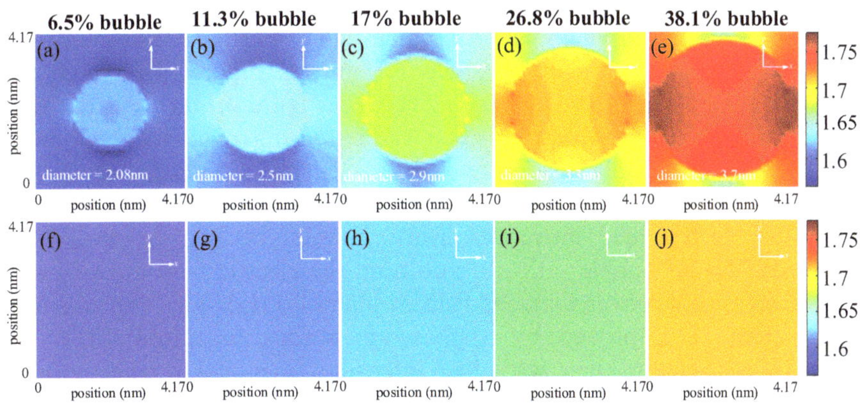

Figure 4.10. Calculated electric field intensities above the mirror (x–y-plane), in which (a)–(e) are scattering results from rigorous FDTD calculations and (f)–(j) are 0th order results from efficient FDTD calculations.

figures 4.10(a)–(e) that stronger electric field intensities can be observed and mostly concentrated in bubbles. This observation provides the information concerning where the high electric flux will be located in the bubble. Moreover, the electric flux within the bubble can be intensified with the increment of the bubble size. For the results from efficient FDTD calculations as shown in figures 4.10(f)–(j), no scattering effect can be observed. Instead, only 0th order images can be found, since the equivalent ML

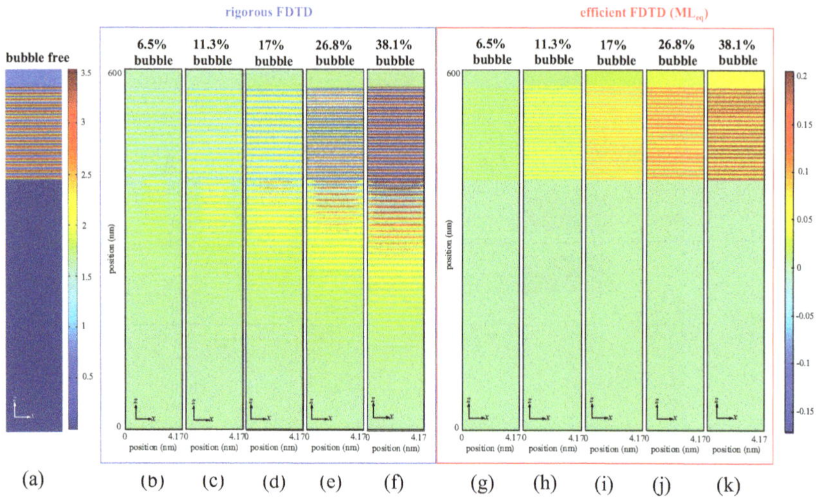

Figure 4.11. Calculated electric field intensities at the center of the *y*-axis (*x*–*z*-plane), in which (a) stands for the results from rigorous FDTD calculation for a bubble-free mirror. (b)–(f) are differences between (a) and rigorous FDTD calculations for 40 Mo/Si (porous) ML based mirrors. (f)–(j) are differences between (a) and efficient FDTD calculations for equivalent 40 Mo/Si (porous) ML based mirrors.

is homogeneous along all directions. Nevertheless, the overall electric intensity is increased with the increment of the bubble ratio. This is the same phenomenon that both the rigorous FDTD method and efficient FDTD method have in common.

The calculated results at the center of the *y*-axis (*x*–*z*-plane) corresponding to figure 4.10 are plotted in figure 4.11, in which figure 4.11(a) is the result from the rigorous FDTD calculation for a bubble-free mirror. Figures 4.11(b)–(f) are the differences of the electric field intensities between figure 4.11(a) and the rigorous FDTD calculation for 40 Mo/Si (porous) ML based mirrors. The phenomenon in figures 4.11(b)–(f) is similar to figures 4.10(a)–(e), the differences of the electric intensities are focused within bubbles. Besides, these figures exhibit that the differences of the electric field intensities inside bubbles become recognizable with the increment of the bubble ratio. In addition, the electric field intensities above the mirror are also stronger than the field in figure 4.11(a). This explains how the reflectivity can be improved by applying a porous Si mixed EUV mirror. Figures 4.10(a)–(e) tell us the electric field intensities (identical to reflectivity) above the mirror of a porous Si mixed EUV mirror can be strengthened if we can increase the bubble ratio or size. On the other hand, no focused electric field intensity inside the mirror can be observed for efficient FDTD calculations owing to homogeneity of the equivalent ML, as shown in figures 4.11(g)–(k). This skin effect remains the same for different bubble ratios. Anyhow, the field strengthen effect above the mirror that can also be observed in rigorous FDTD calculations appears in efficient FDTD calculations. As with rigorous FDTD calculations, figures 4.11(g)–(k) tell us the electric field intensities above the mirror are strengthened with the increment of the bubble ratio or size.

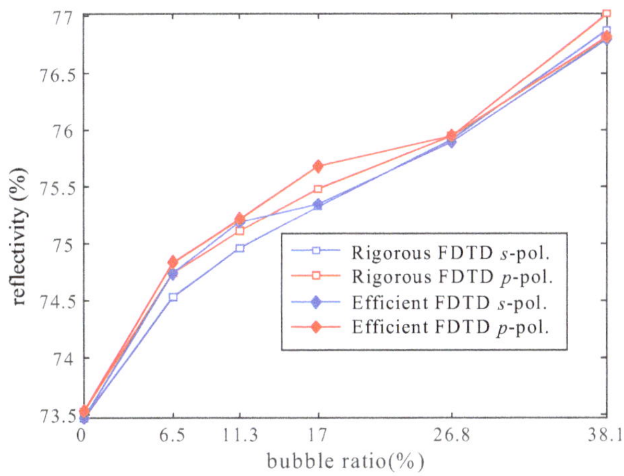

Figure 4.12. Calculated reflectivity for a 40 Mo/Si (porous) ML based mirror under different bubble ratios. Blue and red hollow squares are the results from rigorous FDTD calculations for *s*- and *p*-polarizations. Blue and red solid diamonds are the results from efficient FDTD calculations for *s*- and *p*-polarizations.

Table 4.4. Computational efficiency comparisons between 3D efficient FDTD and rigorous FDTD for a porous Si mixed mirror under EUV incident obliquely at angle of incidence of 6°.

Methods		Rigorous FDTD					Efficient FDTD				
Bubble ratio (%)		6.5	11.3	17	26.8	38.1	6.5	11.3	17	26.8	38.1
CPU time (s)	*s*-pol.	1254	1226	1264	1325	1236	538	591	615	588	595
	p-pol.	1304	1293	1273	1248	1209	542	570	602	604	594

The reflectivity calculated supports the field strengthen effect above the mirror as observed in figure 4.11. The reflectivities calculated from both efficient FDTD and rigorous FDTD calculations for a 40 Mo/Si (porous) ML stack based EUV mirror under different bubble ratios are plotted in figure 4.12. It can be seen that the reflectivities for both calculations are increased with the increment of the bubble ratio, meaning the reflectivity is still predictable using efficient FDTD calculation.

The computational efficiency comparisons between 3D rigorous FDTD and efficient FDTD calculations for a porous Si mixed mirror are provided in table 4.4. From the table, what we see is no matter which bubble ratio were applied, efficient FDTD calculations are almost two times faster than rigorous FDTD calculations. This conclusion can also be observed in sections 4.1 and 4.2. A few summaries for section 4.3:

– From the scattering effect perspective, the efficient FDTD calculation is not able to provide scattering results identical to rigorous FDTD calculations for a porous ML based mirror since the equivalent layer is homogeneous in all directions.

- From the reflectivity estimation perspective, the efficient FDTD calculations provide comparable results to rigorous calculations.
- To sum up, the efficient FDTD calculations can be an efficient tool to estimate reflectivity for a porous ML based EUV mirror based on full 3D calculations.

References

[1] Kim E J, You J H, Kim S S, Cho H K, Ahn J, An I and Oh H K 2010 Influence of mask surface roughness on 22 nm node extreme ultra-violet lithography *Proc. SPIE* **7636** 76362K

[2] Endo A *et al* 2006 Development status of HVM laser produced plasma EUV light source *Int. EUV Symp.* (Barcelona, Spain)*(Oct., 2006)* 1–19

[3] Du Y, Choi C J, Zhang G, Park S J, Yan P Y and Baik K H 2006 TaN-based EUV mask absorber etch study *Proc. SPIE* **6283** 62833D

[4] Chou Y C, Nestinger S S and Cheng H H 2010 CH MPI: Interpretive parallel computing in C *IEEE J. Comput. Sci. Eng.* **12** 54–67

[5] Brewster's law from Wikipedia (https://en.wikipedia.org/wiki/Brewster%27s_angle).

[6] Chung C K, Tsai M Q, Tsai P H and Lee C 2004 Fabrication and characterization of amorphous Si films by PECVD for MEMS *J. Micromech. Microeng.* **15** 136–42

[7] Yang W S, Cho S M, Ryu H J, Cheon S H, Yu B G and Choi C A 2011 Method for depositing amorphous silicon thin film by chemical vapor deposition *US Patent Specification* US20110159669A1

[8] Hui F, Li B, He P, Hu J and Fang Y 2009 Electrochemical fabrication of nanoporous polypyrrole film on HOPG using nanobubbles as templates *Electrochem. Commun.* **11** 639–42

[9] Jamois C, Li C, Gerelli E and Orobtchouk R 2011 *Biosensors-Emerging Materials and Applications* ed P A Serra (Rijeka: InTech) ch 14 New concepts of integrated photonic biosensors based on porous silicon

[10] Suresh K, Ohnuma M, Oba Y, Kishomoto N, Das P and Chini T K 2010 Probing Ar ion induced nanocavities/bubbles in silicon by small-angle X-ray scattering *J. Appl. Phys.* **107** 073504

[11] Lee Y M, Li J H and Tsai K Y 2013 Void-based photonic crystal mirror with high reflectivity and low dissipation for extreme-ultraviolet radiation *J. Micro/Nanolith. MEMS MOEMS* **12** 043005

Chapter 5

Summaries and challenges

Summaries of this book are listed in section 5.1. Challenges ahead of extreme ultraviolet (EUV) mirror design and EUV lithography are provided in section 5.2.

5.1 Summaries

Basic knowledge of three critical areas are the major topics of this book: EUV mirror design using the efficient finite-difference time-domain (FDTD) approach are introduced in chapters 1–3, they are the concepts of photolithographic technologies including EUV lithography, compositions and the physics of EUV mirrors, and the framework of the FDTD method.

In chapter 1, we introduced not only the operational principle of generic photolithography but also described in detail its interrelation with the integrated circuits (IC) fabrication process. Both manufacturing flows were described in the same chapter. Also, EUV lithography that utilizes a completely different physics from generic photolithography was introduced at the end of chapter 1. In general, chapter 1 gave readers an idea of linking IC manufacturing with EUV lithography.

In chapter 2, the generic EUV mirror used in EUV lithography was introduced. We addressed how it is composed and how it can be manufactured. Then, we struck at the heart of this chapter to introduce the types and physics of the multilayer (ML) based EUV mirror. Based on the introduced physics, we presented how the reflectivity of the ML based mirror can be quickly assessed through a 1D analytical model. Afterward, the concept of Bragg Mirror, which is a common mirror design used in EUV applications, was mentioned. In addition, we compared the reflectivity performance between the mainstream design with the Bragg Mirror using a 1D analytical model. At the end of the chapter, a photonic crystal (PHC) based EUV mirror, which has the potential to eliminate/change the loss tangent in materials is indicated. We demonstrated the potential high reflectivity of the PHC based EUV mirror through the 1D analytical model. To sum-up, chapter 2 focused on the working principles of the EUV mirror.

doi:10.1088/978-0-7503-2652-0ch5

In chapter 3, the efficient FDTD method was introduced. We firstly reviewed numerical methods for EUV mirror design, where the advantages to use the FDTD method on EUV mirror design were listed. Then, the computational scheme of rigorous FDTD was described. At that point, we demonstrated how to estimate the required computational resource before starting FDTD computation. This is to avoid potential computational time (CPU) crash or long waiting time. By utilizing a few simple examples, the results indicated that both computational time and required memories can be well-predicted. Furthermore, boundary conditions, which are essential to FDTD computation, were introduced. Both Berenger's perfect matched layer (PML) and periodic boundary condition, which are commonly used in FDTD calculations, were presented in detail. They are also used in our simulation cases in chapter 4. Moreover, the efficient FDTD method, which treats ML structure as one equivalent layer, was presented in section 3.4. Through a benchmark example, the benefit we see in efficient FDTD calculation is to have half the computational time while preserving the numerical accuracy. In short, chapter 3 showed the reader how the efficient FDTD method was constructed and what are the benefits of using this method.

In chapter 4, one 2D case and two 3D cases relative to EUV mirror design were used to demonstrate the efficient FDTD calculations. All the cases were benchmarked with rigorous FDTD calculations. The 2D case concluded that the efficient FDTD method works better than the rigorous FDTD method in consideration of CPU time. Concerning numerical accuracy of the efficient FDTD calculation, the root mean square error (RMSE) of the near-field scattering result between rigorous FDTD calculations and efficient FDTD calculations stayed within 0.26%, which is remarkable. We also justified the reflectivity calculations from rigorous and efficient FDTD methods are very close since the maximum RMSE above the mirror is below 0.1%. The near-field scattering calculations of the first 3D case (mirror patterned with contact holes) indicating that no matter which meshes were applied, efficient FDTD calculations are almost two times faster than rigorous FDTD calculations. Also, no matter which FDTD method was applied, the CPU time of the rough mesh is always shorter than the CPU time of the fine mesh. Concerning mirror reflectivity from the far-field calculations, we observed the reflectivities estimated from Fresnel's equations based solver are always higher than the reflectivities from FDTD calculations owing to model inconsistency. The solver assumes flat layers are calculated thus ignores the geometry of the contact holes. Therefore, the estimated reflectivities curves are relatively smooth compared to the reflectivites calculated from FDTD methods. Regardless, the maximum differences are within 2%. For these cases we concluded the efficient FDTD method can be a useful tool to accelerate the calculations with reasonable numerical accuracy. Regarding the last 3D case (porous ML based mirror), near-field scattering results above the mirror and in the center of the x–z-plane are incomparable between rigorous FDTD calculations and efficient FDTD calculations since the equivalent ML is homogeneous along all directions. The efficient FDTD method is not able to provide scattering results identical to the rigorous FDTD method for a porous ML based mirror. On the other hand, from the reflectivity estimation perspective, the efficient FDTD calculations provided comparable results to rigorous calculations.

5.2 Challenges

Owing to the booming demand of the 5G market, the request for making smaller ICs is raised. Although EUV lithography has been chosen to enable the continued extension of Moore's law after immersion lithography, it still has limits. At this moment the major challenge to all of us could be there is a clear roadmap for the next generation IC manufacturing after EUV lithography. How to keep Moore's law moving on is one major challenge. Also, the computational capability of the existing CPUs and the scattering performance of the efficient FDTD are both challenges. From the computational capability perspective, although the efficient FDTD method present in this book is capable of computing the problem larger than using the rigorous FDTD method, the potential memory overflow and long CPU time issues are still there once the IC size keeps shrinking down or dealing with more complex problems. Seeking more powerful computational hardware to support this research, such as multi-GPU or quantum CPU is another challenge. From the view point of efficient FDTD scattering performance, the efficient FDTD calculation is not able to provide accurate scattering results for buried substance. Thus, the efficient numerical methodology with accurate scattering results is also one of challenges.

www.ingramcontent.com/pod-product-compliance
Ingram Content Group UK Ltd.
Pitfield, Milton Keynes, MK11 3LW, UK
UKHW051849100926
8556IPUK00005B/528

9 780750 326506